Lab Manual to Accompany
DC/AC Circuits and Electronics:
Principles & Applications

Kevin Taylor

Terrence O'Connor

Robert J. Herrick

THOMSON

DELMAR LEARNING

Australia   Canada   Mexico   Singapore   Spain   United Kingdom   United States

THOMSON

DELMAR LEARNING

Lab Manual to Accompany
DC/AC Circuits & Electronics: Principles & Applications

By Kevin Taylor, Terrence O'Connor, and Robert J. Herrick

Vice President, Technology and Trades SBU:
Alar Elken

Editorial Director:
Sandy Clark

Senior Acquisitions Editor:
Dave Garza

Senior Development Editor:
Michelle Ruelos Cannistraci

Marketing Director:
Cynthia Eichelman

Channel Manager:
Fair Huntoon

Marketing Coordinator:
Brian McGrath

Production Director:
Mary Ellen Black

Production Manager:
Larry Main

Senior Project Editor:
Christopher Chien

Art/Design Coordinator:
Francis Hogan

Technology Project Specialist:
Kevin Smith

Senior Editorial Assistant:
Dawn Daugherty

Library of Congress Cataloging-in-Publication Data:

ISBN: 1-4018-8040-1

NOTICE TO THE READER

Contents

Preface

This laboratory manual was designed to accompany the text *DC/AC Circuits and Electronics: Principles & Applications* by Robert J. Herrick and is intended for use in an introductory course in DC and AC circuit analysis where electronics are introduced. The combination of text and laboratory manual is appropriate for an Electrical Engineering Technology or Electrical Technology curriculum and could be effectively used in a survey course for non-majors in an engineering curriculum.

The text and laboratory manual take a non-traditional approach to AC and DC circuits by introducing semiconductor devices such as diodes, transistors, and operational amplifiers as applications of the basic laws studied in circuit analysis. This synergism not only makes circuit analysis more interesting, but also by combining electronics with circuit analysis it better prepares the student for future coursework. Preliminary use of the text and laboratory manual indicates an improvement in student retention. Use of more traditional laboratory exercises (typically employing only resistive circuits and voltage sources) would fail to take advantage of the benefits of the new approach.

Each exercise features the following elements:

- **Text Reference.** This highlights a section or chapter from the text covering the topic of the exercise.

- **Required Equipment and Parts.** The equipment required is typical of that found in most introductory electronics laboratories.

- **Objective.** This outlines the procedures to be completed and may include a brief summary of the concepts under investigation.

- **Pre-lab Requirements.** In order to be better prepared for the laboratory discovery, students are required to make some preliminary calculations; these include entering sample calculations. In many cases, students must build one or two circuits prior to the scheduled laboratory period. By requiring students to prepare for the exercise, we believe that they will understand the task before them and better comprehend the concepts being demonstrated. So that students may identify them easily, pre-lab requirements are highlighted by a gray screen.

- **Performance Checklist.** This is a list of required instructor sign-off steps. It is located in the beginning of the exercise so that students can confirm that they have obtained all of the proper signatures. Some instructors may prefer to put the signatures here rather than in the procedure.

- **Instructor Sign-Offs.** An optional feature incorporated in this lab manual is the use of instructor sign-offs designated by the icon: 🛑. Many instructors inspect and sign off laboratory data at the end of the exercise period after a student has completed all measurements. Unfortunately, errors found here do little to help the student while he or she is taking measurements. In some cases, a misconception may be reinforced due to an incorrect measurement. The instructor sign-offs used in this manual are checkpoints of student comprehension. In addition to providing the instructor with instant feedback about student understanding, this time can be used to have individualized discussions about concepts that students find challenging.

- **Procedures.** These are step-by-step instructions with figures and tables necessary to complete the measurements. In earlier exercises, we assume that the student has limited experience with laboratory equipment, so we provide more detail in the steps. As the student progresses, fewer details are provided in the procedure steps. Some procedures will require an instructor sign-off. Procedures requiring calculations may have space provided where students will show samples of those calculations.

- **Observations.** Here, the students are asked to analyze the measurements made during the procedure. It is recommended that the students complete the observations before continuing on to the next procedure.

- **Synthesis.** Most exercises culminate with a Synthesis section. This section has one or more paragraphs describing what measurements should be taken. The Synthesis steps have less detail than a procedure, but students should be able to build upon techniques learned from earlier procedures to complete the task. Many of the exercises in the latter portion of the manual require computer simulation or spreadsheet work following completion of the measurements.

The text and lab manual are intended for use in either a two-semester or three-quarter course sequence. Each exercise can be completed within a three-hour period. For each semester, we perform thirteen exercises and hold two laboratory practical examinations.

About the Authors

Kevin D. Taylor is an Associate Professor in the Department of Electrical and Computer Engineering Technology at Purdue University–Kokomo. He holds a B.S. in Electrical Engineering from Iowa State University and an M.S. in Electrical Engineering from Southern Methodist University. He has over fifteen years of teaching experience and has been teaching introductory circuits courses for the past several years. Prior to his academic career, he was an IC Design Engineer at Texas Instruments, Inc. Mr. Taylor currently serves as a commissioner for the Technology Accreditation Commission of ABET representing SAE. He is a member of the Engineering Technology Division of ASEE, a member of IEEE, and a registered professional engineer (Pennsylvania).

Terrence P. O'Connor is an Associate Professor in the Department of Electrical and Computer Engineering Technology at Purdue University–New Albany. Professor O'Connor holds an M.S. in Engineering Technology from West Texas State University and B.S. in Engineering Technology from Northern Arizona University. Mr. O'Connor has taught for twenty years in electronics and engineering technology. His engineering experience was in the aerospace industry with Lockheed Martin as a test engineer in the microelectronics laboratory

Robert J. Herrick is the Robert A. Hoffer Distinguished Professor of Electrical Engineering Technology at Purdue University–West Lafayette and is currently serving as head of the Electrical and Computer Engineering Technology Department. He has received numerous teaching awards including HP Outstanding Laboratory Instruction Award, the ASEE Illinois-Indiana Outstanding Teaching Award, the Purdue University Murphy Award for Outstanding Undergraduate Teaching, the School of Technology Dwyer Award for Outstanding Unvergraduate Teaching, and the EET department CTS Electronics Award for Outstanding Undergraduate Teaching. Prior to his academic career, he worked in design and development of electronic switching systems at Bell Telephone Laboratories and ITT Advanced International Technology Center and also served as an engineering consultant for several major industries in the areas of controls, instrumentation, technical publications, and computer software design and development. He served as an electronics technician of ground-based navigational aid systems in the United States Air Force.

Acknowledgments

The authors and Delmar Learning would like to thank the following reviewers for their input and suggestions:

G. Thomas Bellarmine, Florida A & M University

John Blankenship, DeVry University

Yolanda Guran, Oregon Institute of Technology

William Hirst, DeVry University

Sang Lee, DeVry University

The authors would like to thank the following people for their help, guidance, support, and suggestions in writing this laboratory manual: Alan Rainey, Professor Theodore Fahlsing, Professor Athula Kulatunga, Roger Davis, Professor J. Michael Jacob, Professor Robert Hofinger, Stan Dick, Gitte Strauss, Professor Russ Aubrey, and the many students who "test drove" these laboratory exercises. We would also like to thank Michelle Ruelos Cannistraci and Greg Clayton at Delmar, a Thomson Learning company, for their suggestions and patience in helping us to complete this project.

Kevin Taylor would like to thank his wife Ann and daughters for putting up with the long hours in front of the computer, and his parents for their support and nurturing.

Terrence O'Connor would like to thank his wife Becky for her support in this effort.

Robert J. Herrick would like to thank Professors Taylor and O'Connor for working diligently to author this accompanying laboratory manual; and he would especially like to thank his wife Becky for letting him work many long hours creating this very unique approach that enhances student understanding with increased enjoyment in learning electronics engineering technology.

Comments or suggestions about how to improve the lab manual are appreciated. Please send them to:

Kevin D. Taylor
Purdue University at Kokomo
Kelly Center, Room KC265A
P.O. Box 9003
Kokomo, IN 46904-9002
k.d.taylor@ieee.org

Kevin D. Taylor
Terrence P. O'Connor
Robert J. Herrick

Component Suppliers

Learning Systems, Inc.

531 Carrolton Blvd.
W. Lafayette, IN 47906-2335
Phone/Fax: (765) 497-6447
Email: info@learningsystemsinc.com
http://www.learningsystemsinc.com
Parts kit number EET107-157

Martek Electronics

3729 White Chapel Way,
Raleigh, NC 27615
Phone: (888) 223-5487
Fax: (919) 518-1458
Email: MartekElec@aol.com
www.MartekElec.com
Parts kit number EET107-157

Abra Electronics Corporation

1320 Route 9
Champlain, NY 12919
Phone/Fax: (800) 717-2272
Email: sales@abra-electronics.com
www.abra-electronics.com
Parts kit number EET107-157

Instruments and Measurements

Name: _____   Date: _____

Lab Section: _____ _____   Lab Instructor: _____
                  day   time

Text Reference

DC/AC Circuits and Electronics: Principles and Applications
Chapter 2: Current, Voltage, and Common

Materials Required

Triple power supply (2 @ 0–20 volts DC; 1 @ 5 volts DC or similar)
5 Assorted batteries (preferably of different type labeled A through E)
1 Analog Volt-Ohm-Meter (VOM) (*optional*)
1 Analog milliammeter
1 Digital multimeter
1 Resistor substitution box

Introduction

Effective use of laboratory instruments to make measurements is critical to success in the electronics field. This exercise introduces several concepts.

In this exercise, you will:

- Use a battery and a power supply to create a voltage
- Measure a voltage using a voltmeter
- Combine two voltages to create a larger voltage
- Measure resistance using a resistor substitution box
- Measure current in a circuit.

Pre-Lab Activity Checklist

Pre-Lab

☐ Read the laboratory exercise.

☐ Review text Chapter 2 on voltage, current, and proper connection of voltmeters and ammeters. Have the text available for reference during the laboratory period.

Performance Checklist (An instructor's signature is required for each.) ■

☐ Table 1-2   Demonstrate 19-V, –8-V, and 27-V readings.
☐ Table 1-3   Demonstrate 100-Ω and 220-Ω measurements.
☐ Table 1-4   Demonstrate current measurement using the analog ammeter.
☐ Table 1-4   Demonstrate current measurement using the DMM ammeter.

Procedure 1-1

Battery Measurement

 WARNING!   Placing a battery in your mouth may result in extreme burns. Connecting a battery in a manner other than instructed to do so may result in a battery explosion.

1. Obtain the collection of batteries from your instructor. The type and nominal voltage of each battery should be imprinted on each battery (AA or D cell; 1.5 V or 9 V, for example). Enter that information in Table 1-1.

TABLE 1-1   Battery Voltage Measurements

Battery Label	Battery Type	Voltage Nominal	Voltage VOM*	Voltage DMM	Percent Difference
A →		V	V	V	%
B →		V	V	V	%
C →		V	V	V	%
D →		V	V	V	%
E →		V	V	V	%

Note: Always Include Units

V for volt

 WARNING!   Never apply voltage to a meter when it is in the OHMS or CURRENT position. Randomly switching through the meter control modes (volts, amps, ohms) with the meter connected to a live circuit is a bad practice that could damage the meter and/or the circuit.

2. Measure each battery using the analog Volt-Ohm-Meter (VOM) *if available* and enter that information in Table 1-1. Start with the meter set on the highest voltage scale. Place the red lead on the positive (+) terminal and the black lead on the negative (−) terminal, as shown in Figure 1-1, then reduce the voltage scale setting to the one just above the voltage being measured.

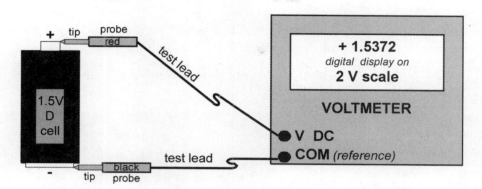

FIGURE I-I Example of Battery Voltage Measurement

◨ Note: Used batteries may measure substantially lower than rated (nominal) value.

3. Return the VOM to the highest scale. Using one of your batteries, *briefly* measure the battery voltage with the meter leads reversed (black on +; red on −) and observe the meter action. **Avoid this practice on analog meters!**

4. Measure each battery using the digital multimeter (DMM) and enter that information in Table 1-1. Place the red lead on the positive (+) terminal and the black lead on the negative (−) terminal, as shown in Figure 1-1.

5. Using the DMM and one of your batteries, measure the battery voltage with the meter leads reversed (black on +; red on −) and observe the meter action. Note that when using the DMM, reversing the leads is okay.

SAMPLE CALCULATIONS

Sample Calculations are required steps where you show samples of any calculations needed. This provides a record for later reference on how you made a calculation.

The format for all calculations is shown below. WHAT is the variable you are solving for. FORMULA shows the formula using variables. SUBSTITUTED VALUES is the formula with numbers substituted for the variables. RESULT is the final answer. Here is an example:

$$\underset{\downarrow}{\text{WHAT}} \quad = \quad \underset{\downarrow}{\text{FORMULA}} \quad = \quad \underset{\downarrow}{\text{SUBSTITUTED VALUES}} \quad = \quad \underset{\downarrow}{\text{RESULT}}$$

$$\% \text{ difference} = \frac{V_{NOM} - V_{DMM}}{V_{DMM}} \times 100\% = \frac{1.55\,V - 1.5\,V}{1.5\,V} \times 100\% = 3.23\%$$

Show the sample calculations for the percent difference of the voltage measurement of battery A:

$$\underset{\downarrow}{\text{WHAT}} \quad = \quad \underset{\downarrow}{\text{FORMULA}} \quad = \quad \underset{\downarrow}{\text{SUBSTITUTED VALUES}} \quad = \quad \underset{\downarrow}{\text{RESULT}}$$

$$\% \text{ difference} = \frac{V_{NOM} - V_{DMM}}{V_{DMM}} \times 100\% = \frac{\underline{\quad}\,V - \underline{\quad}\,V}{\underline{\quad}\,V} \times 100\% = \underline{\quad}\%$$

Observations   ◼

1. When the VOM was connected with the black lead on the positive (+) terminal and the red lead on the negative (–) terminal, what was the effect?

_____

2. When the DMM was connected with the black lead on the positive (+) terminal and the red lead on the negative (–) terminal, what was the effect?

_____

Procedure 1-2
Power Supply Use and Voltage Measurement

1. Locate the power supply on the laboratory bench and determine the type of supply from the list below. Review text Section 2.8 if you are not sure.
   ☐ Triple independent DC power supply with separate common (text Figure 2-31)
   ☐ Triple independent DC power supply with earth common (text Figure 2-32)
   ☐ Triple DC power supply with internal analog common (text Figure 2-33)
   ☐ Other (specify) _____

2. Turn on the DC supply and, using the meter on the supply (if available), set one of the 0- to 20-V output voltages to 3 V.

3. Using the DMM, measure that 3-V output and record the value in Table 1-2.

4. Leave the power supply on, readjust that same output for 10 V (using the supply's meter) and read and record the DMM voltage measurement in Table 1-2.

TABLE 1-2 Power Supply (PS) Voltage Measurements

Power Supply Voltage*	DMM Voltage	% Difference**
3 V		
10 V		
19 V	STOP *demonstrate*	
–8 V	STOP *demonstrate*	
27 V	STOP *demonstrate*	

* Read from meter on supply
** Use same format as shown in step 8.

Note: **Always Include Units**
V for volt

5. Repeat the previous step with the power supply meter set to 19 V. Read and record the DMM voltage measurement in Table 1-2.

6. **Do not alter the 19-V output!** Switch your power supply meter to read a different 0- to 20-V output, and adjust that output for –8 V. If your power supply only provides positive voltages, you will need to reverse the DMM leads. Read and record the DMM voltage measurement in Table 1-2.

7. Using a jumper, connect the positive (+) terminal of the 8-V output to the negative (–) terminal of the 19-V output.

8. Using the DMM, measure the voltage from the positive (+) terminal of the 19-V output to the negative (–) terminal of the 8-V output. The difference should be 27 V (19 V – (–8 V) = 27 V). If you have difficulty here, consult with your instructor.

9. Demonstrate the 19-V, –8-V and 27-V readings and have your instructor sign below.

🛑 Instructor sign-off of 19-V, –8-V, and 27-V readings in Table 1-2 _____

SAMPLE CALCULATIONS

Show sample calculations below using the format described in Procedure 1-1.

$$\% \text{ difference} = \frac{V_{\text{DMM}} - V_{\text{PS}}}{V_{\text{PS}}} \times 100\% =$$

Procedure 1-3

Resistance Measurement

Using the Ohmmeter

The ohmmeter is the electronics test instrument that measures resistance. Two very important rules are:

☐ **ALL POWER MUST BE REMOVED** from the circuit when making resistance measurements to prevent damage to the meter and/or false readings.

☐ At least **one end** of the component being measured must be **DISCONNECTED** from the rest of any electronics circuit to prevent extraneous parallel resistance readings.

A representation of an ohmmeter is shown in Figure 1-2.

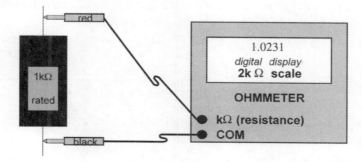

FIGURE 1-2 Measuring Resistance with an Ohmmeter (Test leads can be reversed)

1. Switch the DMM to read resistance (ohms = Ω).

2. Measure the resistance of a wire and record it in Table 1-3.

3. Measure the resistance of yourself using a loose grip on the DMM leads. Repeat the measurement while you squeeze the probe tips tightly. Record the measurements in Table 1-3.

4. Measure the resistance of an open circuit (air) and record it in Table 1-3.

Table I-3 Resistance Measurements

Resistance	Nominal Value	Resistance Measured	Percent Difference
R_{wire}	(very low ≈ 0)		
$R_{loose\ grip}$	(2000-kΩ scale)		(may vary)
$R_{tight\ grip}$	(2000-kΩ scale)		(may vary)
R_{air}	(20-MΩ scale)		(large value)
Substitution Box:	47 Ω		
	100 Ω	🛑	demonstrate
	220 Ω	🛑	demonstrate
	470 Ω		
	1000 Ω or 1 kΩ		
	2.2 kΩ		
	4.7 kΩ		

Note: Always Include Units

Ω for ohms
kΩ for 10^3 ohms
MΩ for 10^6 ohms

5. Obtain a resistance substitution box (or decade box) and adjust it to 47 Ω. Using the lowest meter scale that produces a reading, measure and record the resistance of the substitution box in Table 1-3. Resistor resistance does not have polarity so you may connect the leads either way.

6. Reverse the meter leads and observe the result.

7. Measure and record in Table 1-3 the resistance of the substitution box for settings of 100 Ω and 220 Ω. Demonstrate these to your instructor and have the instructor sign off.

8. Measure and record in Table 1-3 the resistance of the substitution box for settings of 470 Ω, 1 kΩ, 2.2 kΩ, and 4.7 kΩ. Always use the most sensitive range for the most accurate reading.

9. Calculate the percent difference for the 1-kΩ resistance and enter the value in Table 1-3. Using the required format, show your calculation in the space provided.

🛑 Instructor sign-off of 100-Ω and 220-Ω readings in Table 1-3 _____

SAMPLE CALCULATIONS

Show a sample calculation of the % difference in the 1-kΩ reading.

$$1\ k\Omega \quad \%\ \text{difference} = \frac{R_{\text{MEASURED}} - R_{\text{NOMINAL}}}{R_{\text{NOMINAL}}} \times 100\% = \qquad\qquad = \underline{\qquad}\ \%$$

Observations

1. What was the effect of reversing the leads when measuring the 47-Ω resistance?

    _____

2. If the meter reads out of range, does that mean the resistance is infinite?   ☐ Yes   ☐ No
    Explain your answer.

    _____

    _____

Procedure I-4

Current Measurements

1. Using the DMM as a DC voltmeter, adjust one output of the DC power supply to deliver +5 V. Start at 0 V and raise the voltage to +5 V. Then turn off the power supply, but do not change the setting.
2. Construct the circuit shown in Figure 1-3, using the highest current scale available. (Your analog current meter may differ from Figure 1-3.) Be sure you make *metal-to-metal* contact when making connections. Conventional current (*I*) flows from the positive (+) terminal to COM. **Do not turn the power supply on until instructed to do so.**

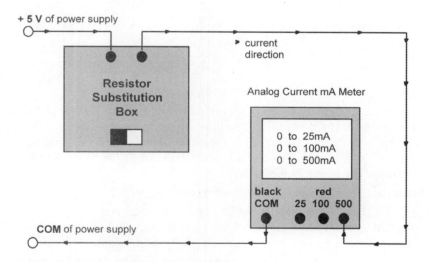

FIGURE I-3   Milliammeter Analog Current Measurement

 WARNING!   Always **break the circuit** and **insert the current meter** to reconnect the break. The current meter has very low resistance so it acts like a **SHORT**.

NEVER connect the current meter like a voltmeter. If the current meter is placed across a voltage source or across a voltage drop, the current meter's *short-circuit nature* will draw *a large current* through the meter. If the current is excessive, it could cause circuit damage, meter damage, or blow the protective ammeter fuse (if it has one).

3. Set the resistor substitution box to 47 Ω.

4. Turn on the DC supply, then measure and record the circuit current in Table 1-4. Begin by using the highest current scale available and reduce the scale as needed before recording the value. *If the meter indicates beyond full scale* (known as "pegging the meter"), *remove the power immediately, then increase to the appropriate scale.* Using the lowest scale that is not beyond full scale provides the most accurate reading. Demonstrate this to your instructor and obtain a sign-off.

TABLE 1-4 Current Measurements

Resistance Nominal	Current Expected	Current mA Meter	Current DMM	Percent Difference
47 Ω	106 mA	🛑	🛑	*demonstrate*
100 Ω	50 mA			
220 Ω	22.7 mA			
470 Ω	10.6 mA			
1 kΩ	5.0 mA			
2.2 kΩ	2.3 mA			
4.7 kΩ	1.1 mA			

Note: Always Include Units

A for Amps   **mA** for 10^3 A   **μA** for 10^6 A

🛑 Instructor sign-off of current in Table 1-4 using analog ammeter _____

5. Turn off the DC power supply (**do not** adjust the voltage).

6. Repeat the current measurements for resistance settings of: 100 Ω, 220 Ω, 470 Ω, 1 kΩ, 2.2 kΩ, and 4.7 kΩ. **Turn off the power supply while changing the resistance.** Table 1-4 specifies the expected current based on Ohm's Law ($I = V/R$). Be sure to connect the milliammeter to the lowest appropriate range (start initially with the highest range, then reduce as needed). **Again, if the meter is beyond full scale remove the power immediately!**

7. Turn off the DC power supply (**do not** adjust the voltage).

8. Replace the analog current meter with the DMM current meter. To set up the DMM to read current, move the RED DMM test lead to the "mA" connection, select "mA" mode, and then select the proper range. If you are unsure of the range to use, start with the highest range and work your way down.

9. Turn on the power supply (it should still be adjusted for 5 V). Using the DMM current meter, measure and record in Table 1-4 the circuit current for the resistance (ohm) settings of 47 Ω. Demonstrate this to your instructor and have the instructor sign here.

🛑 Instructor sign-off of current in Table 1-4 using a DMM ammeter _____

10. **Turn off the power supply.**

11. Repeat the current readings using the DMM current meter and record in Table 1-4 the circuit current values for the resistance (ohm) settings of: 100 Ω, 220 Ω, 470 Ω, 1 kΩ, 2.2 kΩ, and 4.7 kΩ. **Turn off the power supply while changing the resistance.**

12. When you have finished taking current measurements with the DMM, return the DMM and its cables to the DC *voltmeter mode.* This is to prevent you from accidentally using the current meter as a voltmeter and damaging the meter.

13. Calculate the percent difference for the DMM current and enter the result in Table 1-4. Show the sample calculation below.

SAMPLE CALCULATIONS

$$\% \text{ difference} = \frac{I_{\text{EXPECTED}} - I_{\text{DMM}}}{I_{\text{DMM}}} \times 100\% = \qquad\qquad = \underline{\qquad\qquad} \%$$

Observations ■

1. The voltage was maintained at 5 V. As the resistance was increased, what was the effect on the current?

   _____

2. Which meter was easiest to use?   ☐ Analog VOM   ☐ DMM

   Why? _____

3. Which ammeter scale (analog or DMM) provides the most accurate reading?

                                                    ☐ Analog VOM   ☐ DMM

All remaining laboratory exercises have pre-lab calculations. These calculations must be completed prior to making any measurements. The Sample Calculations are boxed and precede any Observations. Pre-lab checklists are indicated with a marginal tab.

Read the following instructions to prepare your tool kit and protoboard for the remaining exercises.

If you are required to purchase a laboratory parts kit and tools, you must purchase these items and complete the check-off list prior to the beginning of the second laboratory period.

Your Personal Parts Kit Check-Off List   ■

Name: (printed) _____

Sign and date that you have completed the following:

☐ Obtained tools in tool kit list (see page 13)

☐ Matched your parts with the parts kit list (see page 13–14). Reported any shortages to your instructor.

☐ Organized your tool box

☐ Prepared power supply leads (ask for assistance if needed)

☐ Properly wired your Protoboard 10 (see page 15)

Signature _____   Date_____

SUBMIT THIS SHEET PRIOR TO STARTING EXERCISE 2

Tool Kit

The items for your tool kit can be purchased at a hardware store.

1. **Tools**
   - ☐ 1 Tool/Tackle box* (for storing parts and tools)
   - ☐ 1 Pliers, precision needle-nose, 5 inch
   - ☐ 1 Diagonal cutters, miniature, 5 inch
   - ☐ 1 Wire strippers and cutters, 5 inch
   - ☐ 1 Precision screw driver set of 6, 1–3 mm slotted, # 0 and # 1 Phillips

2. **Multimeter (optional)**
   - ☐ 1 Multimeter (Available for under $30. See your instructor for suggestions.)

3. **Soldering Accessories (optional)**
   - ☐ 1 Solder pencil, 25–30 watts
   - ☐ 1 Soldering iron stand
   - ☐ 1 Solder wick
   - ☐ 1 Desoldering tool, vacuum, spring loaded
   - ☐ 1 Solder in dispenser

*Toolbox   Toolbox type and style is the choice of the student. Medium to large plastic fishing-tackle boxes suitable for storing many small parts and tools are a lightweight and economical solution. Students also use small plastic organizers to store components. If you have storage lockers available, consider the locker size before purchasing a toolbox.

Parts Kit

Quantity	Value	Type
Capacitors	(50 W VDC or better)	
☐ 1	0.01 µF	Film
☐ 3	0.1 µF	Film
☐ 1	0.47 µF	Film
☐ 1	1 µF	Electrolytic
☐ 1	10 µF	Electrolytic
☐ 1	22 µF	Electrolytic
☐ 2	47 µF	Electrolytic
☐ 3	100 µF	Electrolytic
☐ 1	470 µF	Electrolytic

	Quantity	Value	Type

Diodes

☐	4	1N4001	Rectifiers
☐	2	1N753 or IN4735	6.2 V Zener
☐	2	Red	LED
☐	2	Green	LED

Fuses

☐	5	1 Ampere	8AG Instrument

Inductors

☐	1	1 mH	
☐	1	22 mH	
☐	1	33 mH	

Integrated Circuits and Semiconductors

☐	1	XR 2206	Function Generator Integrated Circuit
☐	1	2N3055	Transistor, NPN
☐	2	2N3904	Transistors, NPN
☐	1	LM 34DZ	Temperature sensor; Volts/°F
☐	1	LM 340-5	5-volt regulator
☐	3	LM 324	LM 324; Low power Quad Op Amp; 14 pin DIP; packaged on anti-static (pink) foam to protect leads

Potentiometer

☐	1	10 kΩ	Single-turn; trimmer; fits Global Specialties wiring boards
☐	1	10 kΩ	Multi-turn; trimmer; fits Global Specialties wiring boards

Resistor Kit, in Ohms, 1/4 Watt

☐	5	each of 73 standard values, from 1 to 4.7 M, 5%	

Resistors, other

☐	5	1.1 kΩ	1/4 watt; 5%

Other Devices/Items

Quantity	Type
☐ 1	Transducer, sound (MS-3 or equivalent)
☐ 1	Switch, SPDT miniature toggle (ON-OFF-ON)
☐ 1	Black test probe wire, 20 inch
☐ 1	Green test probe wire, 20 inch
☐ 1	Red test probe wire, 20 inch
☐ 2	Red banana plugs
☐ 2	Black banana plugs
☐ 2	Green banana plugs
☐ 1	Banana plug; yellow
☐ 1	Multi-strand 22 AWG wire, 1 foot.
☐ 3	BNC to mini-grabber test probe, 36 inch
☐ 1	Banana plug to mini-grabber test probe, black
☐ 1	Banana plug to mini-grabber test probe, red
☐ 1	Protoboard, PB-10
☐ 1	Wiring board, UBS 100

Optional: Obtain at an electronics supply store

Box of pre-cut 22-gauge jumper wires (various lengths)

A list of component vendors is provided at the end of the Preface on page x.

Wiring the Protoboard 10

Use only solid 22-gauge wire
Refer to Figure 1-4.

- Connect a red wire from the red (left) screw terminal to the top horizontal bus.
- Connect a red jumper across the middle of the top horizontal bus.
- Connect a green wire from the green (center) screw terminal to the bottom horizontal bus.
- Connect a green jumper across the middle of the bottom horizontal bus.
- Connect a black wire from the black (right) screw terminal to the inner top horizontal bus.
- Connect a black jumper across the middle of the inner top horizontal bus.
- Connect a black wire from the black (right) terminal to the inner bottom horizontal bus.
- Connect a black jumper across the middle of the inner bottom horizontal bus.

⊐ Note: When connecting power supply leads to the screw terminals, use the corresponding colored jumpers.

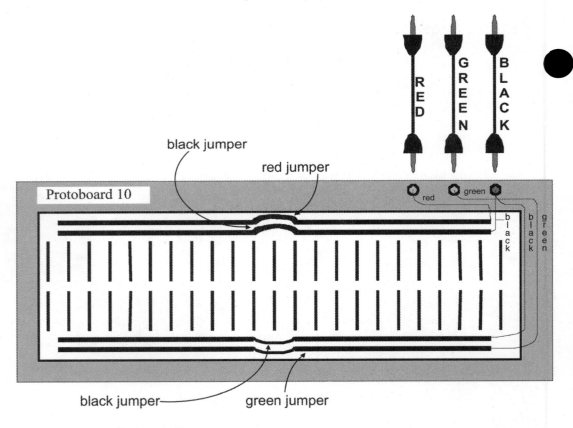

FIGURE 1-4   Protoboard 10 Connections

Connections and Measurements

Name: _____ Date: _____

Lab Section: _____ _____ Lab Instructor: _____
　　　　　　　　　day　　　　　　　time

Text Reference　　　　　　　　　　　　　　　　　　　■

DC/AC Circuits and Electronics: Principles and Applications
Chapter 3: Resistance

Materials Required　　　　　　　　　　　　　　　　■

Triple power supply (2 @ 0–20 volts DC; 1 @ 5 volts DC)

3　1-kΩ resistors

1　470-Ω resistor

1　330-Ω resistor

1　Light-emitting diode (LED), red

1　Digital multimeter

1　USB-100 wiring board

1　Protoboard 10 (pre-wired by student)

Introduction　　　　　　　　　　　　　　　　　　■

In this exercise, you will:

- Learn about the internal connections of the wiring board used in most exercises
- Build a resistive series circuit and examine the voltages and current
- Examine a series circuit using a light-emitting diode (LED) in both forward- and reverse-bias conditions. The LED allows current to flow when it is forward biased, but not when it is reverse biased.
- Plot the current vs. voltage characteristics of a resistor and an LED.

Pre-Lab Activity Checklist　　　　　　　　　　　■

Pre-Lab

☐ Complete the parts kit inventory (if required) at the end of Exercise 1.

☐ Prepare the Protoboard 10 (instructions at the end of Exercise 1).

☐ Prepare your tool kit (instructions at the end of Exercise 1).

☐ Obtain three 1-kΩ resistors and build the circuit shown in Figures 2-4 and 2-5. Resistor leads should be trimmed to about 13 mm (1/2 inch) so that they are slightly off the surface of the wiring board. Connecting wires should have 13 mm (1/2 inch) stripped off of each end and be sized to avoid large loops. Neatness here prevents wiring errors that waste valuable laboratory time.

☐ Review text Chapter 2 on voltage, current, and proper connection of voltmeters and ammeters.

☐ Review text Chapter 3 on resistance and resistance measurement.

☐ Complete Figures 2-9 and 2-10 showing bubble notation and planned board layout. Figures 2-2, 2-3, and 2-4 show examples of bubble notation and proper board layout planning.

Performance Checklist

☐ Pre-lab completed?   (STOP) Instructor sign-off _____

☐ Table 2-3: Demonstrate current and voltage readings.

☐ Table 2-8: Demonstrate V_R and I for 10-V supply voltage.

☐ Attach your signed parts check-off list with this exercise submission.

Procedure 2-1

Wiring Board Continuity

1. Using a USB-100 wiring board (not the Protoboard 10 with the three terminals) and the DMM as an ohmmeter, measure for continuity between the points "a" and "b" in Figure 2-1. You will need to insert a 22-gauge wire into the board. You should measure a low resistance (< 0.1 Ω). Enter your measured value in Table 2-1 as either "0 Ω" for low resistance or "∞ Ω" for an open circuit.

2. Repeat the measurement for the remaining connection points in Table 2-1.

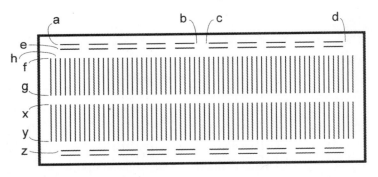

FIGURE 2-1   Wiring Board Layout

TABLE 2-1   Wiring Board Connectivity for Figure 2-1

Connection	a to b	c to d	f to g	x to y	a to e	b to c	e to f	f to h	g to x	y to z
Expected	0 Ω					∞ Ω				
Measured										

Legend: "0 Ω" short (continuity or low resistance)   "∞ Ω" open (no continuity or high resistance)

Procedure 2-2

Measuring Voltage & Current in a Series Circuit

1. Obtain three 1-kΩ (1000-Ω) resistors. Starting with the color band closest to the edge, the order should be brown (1), black (0), red (× 10²) (That is: $10 \times 10^2 = 10 \times 100 = 1000 \ \Omega$).

2. Compare the circuit of Figure 2-2 with the circuit of Figure 2-3. The circuit of Figure 2-3 uses "bubble notation" to replace the voltage source shown in Figure 2-2. The +12 V side of the source voltage (connected to node "a") is replaced with a +12 V bubble. The −12 V side of the source voltage (tied to common) is replaced with a COM bubble. In the future, you will redraw circuits using the bubble notation.

3. Build the circuit of Figure 2-2. The circuit should be neatly laid out like the ones shown in Figures 2-4 and 2-5. The position of components should resemble the schematic placement.

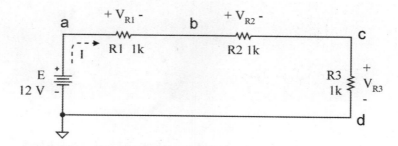

FIGURE 2-2   Circuit for Procedure 2-2

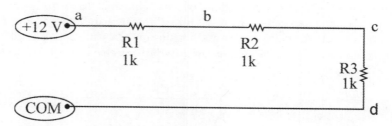

FIGURE 2-3   Circuit for Procedure 2-2 Using Bubble Notation

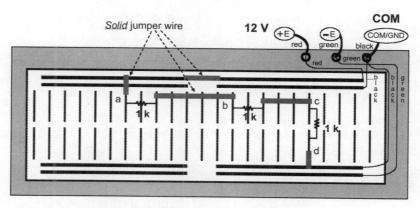

FIGURE 2-4   Wiring Board Layout for Procedure 2-2 (*pre-lab*)

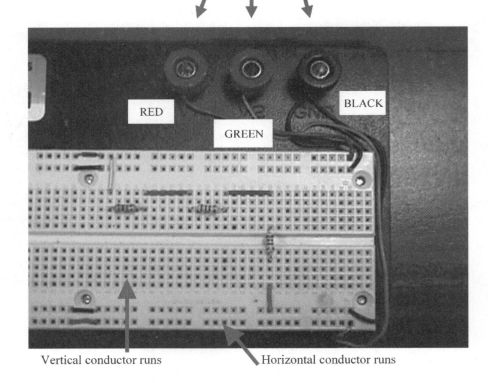

Good metal-to-metal connection.
Do not pinch wire insulation.

RED

GREEN

BLACK

Vertical conductor runs

Horizontal conductor runs

FIGURE 2-5   Photo of Wiring Board Layout for Procedure 2-2 (*pre-lab*)

4. Measure each resistor and enter the value in Table 2-2. To perform this properly you must always disconnect at least one end of the resistor to eliminate parallel resistance paths. Figure 2-6 shows one measurement method where the resistor remains in the wiring board, but the jumper connecting it to the circuit is removed. *When building a circuit, always measure the resistors first to be sure they are the correct value and within tolerance.*

TABLE 2-2   Resistor Data Table for Figure 2-2

Resistor	Nominal Value	Measured Value	Percent Difference	Meets Spec? (within 5%)	
R_1	1 kΩ			☐ Yes	☐ No
R_2	1 kΩ			☐ Yes	☐ No
R_3	1 kΩ			☐ Yes	☐ No

5. Turn ON and adjust one output of the DC supply to 12 V. Use the meter on the DC supply for an approximate setting.

6. Connect the positive (+ red) terminal on the DC supply to the red screw terminal on the Protoboard 10 and the negative (– black) terminal on the DC supply to the black screw terminal on the Protoboard 10.

7. Place your DMM in the voltmeter setting and measure the DC supply voltage. Adjust it to 12 V and record the value (E) in Table 2-3.

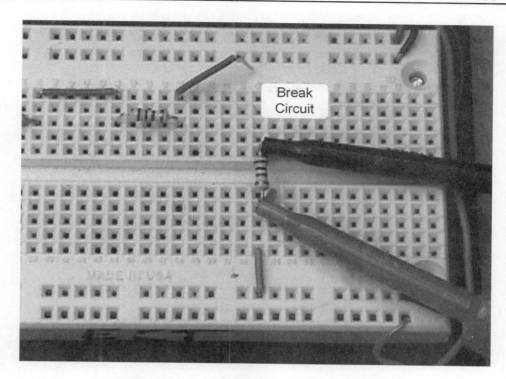

FIGURE 2-6 Proper Resistance Measurement of a Resistor

TABLE 2-3 Data for Procedure 2-2

Variable	Expected Value	Measured Value	Percent Difference
E	12 V		
I	4 mA	🛑	
V_{R1}	4 V		
V_{R2}	4 V		
V_{R3}	4 V		
V_a	12 V		
V_b	8 V		
V_c	4 V	🛑	
V_{ab}	4 V	🛑	
V_{ba}	−4 V		
V_{ac}	8 V		
V_{ca}	−8 V		
V_{ad}	12 V		
V_{da}	−12 V		

8. Turn OFF the DC supply (*but do not change the settings*).
9. Place your DMM in the current meter (ammeter) setting and move the test leads to the ammeter inputs.
10. Break the circuit by lifting one end of a jumper as shown in Figure 2-7 and replacing the open with the current meter. Place the current meter on the highest scale.

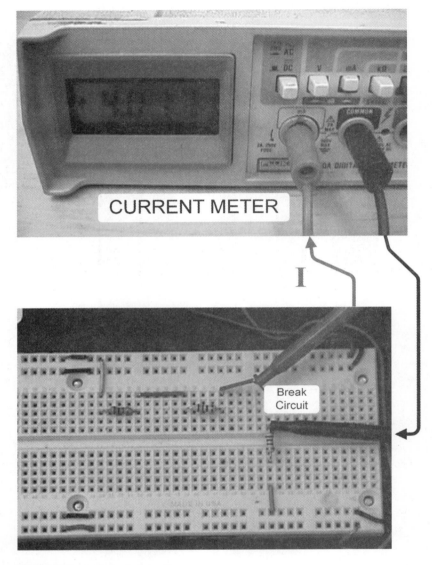

FIGURE 2-7 Resistor Current Measurement

11. Turn ON the DC supply and record the circuit current (*I*) in Table 2-3. Reduce the DMM scale to the lowest scale that produces a reading.
12. Turn OFF the DC supply (*but do not change the settings*) and return your current meter to the voltage setting and lead placement.
13. Turn ON the DC supply and record the remaining voltages listed in Table 2-3 as shown in Figure 2-8. Use the lowest DMM scale that produces a reading.
14. Turn OFF the DC supply. Calculate and record the percent differences in Table 2-3. Show one sample calculation in the space provided below the table.

🛑 Instructor sign-off of the current measurement in Table 2-3 _____

🛑 Instructor sign-off of the voltage measurements in Table 2-3 _____

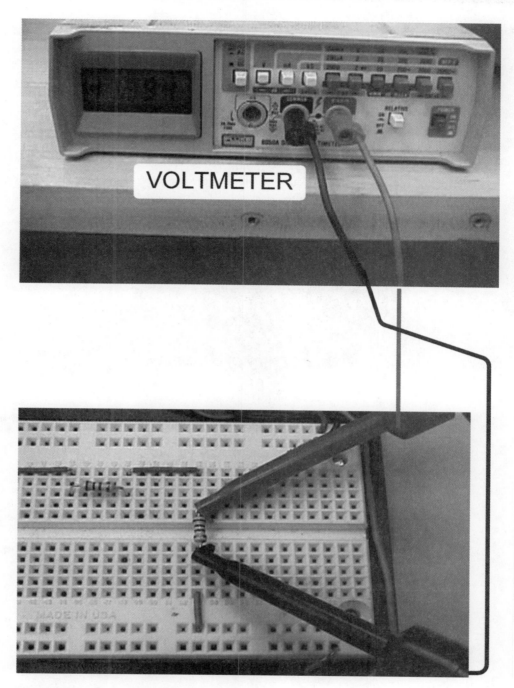

FIGURE 2-8   Resistor Voltage Measurement

SAMPLE CALCULATIONS

Sample calculation of percent difference:

$$\% \text{ Difference} = \frac{\text{Measured Value} - \text{Expected Value}}{\text{Expected Value}} \times 100\% =$$

Observations

1. Why should one measure resistors before using them in a circuit?

_____

2. When taking measurements, why should one use the lowest DMM scale that produces a reading?

_____

_____

3. Are the measured voltages and currents close to the expected values? Quantify your answer in terms of the percent differences calculated.

_____

_____

Procedure 2-3

Series Circuit with *Forward-Biased* LED

1. Redraw the schematic of Figure 2-9, using bubble notation, in the box provided in Figure 2-10. If necessary, refer to the previous procedure.

2. Draw the intended wiring board layout on Figure 2-11.

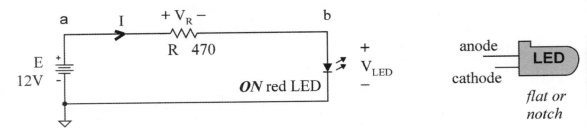

FIGURE 2-9 Series Circuit with a Forward-Biased LED

FIGURE 2-10 Bubble Notation Representation of Figure 2-9 (*pre-lab*)

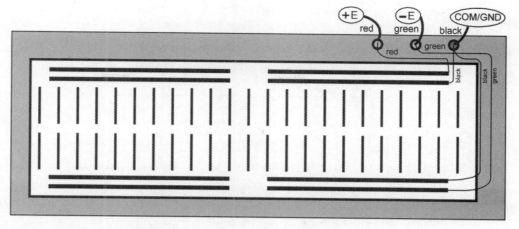

FIGURE 2-11 Intended Wiring Diagram of Circuit of Figure 2-9 (*pre-lab*)

3. Obtain the red LED and the 470-Ω resistor from your parts kit. The resistor color code for a 470-Ω resistor is yellow (4), violet (7), and brown ($\times 10^1$) or ($47 \times 10 = 470\ \Omega$).

4. Build the circuit of Figure 2-9 using proper wiring techniques. Use short leads and a layout similar to the schematic.

5. Measure the resistance of the 470-Ω resistor using proper techniques (open one end, DC supply OFF) and record the value in Table 2-4.

TABLE 2-4 Resistor Data Table for Figure 2-9

Resistor	Nominal Value	Measured Value	Percent Difference	Meets Spec? (within 5%)
R	470 Ω			☐ Yes ☐ No

6. Connect the DC power supply to your circuit and turn the supply ON. Adjust the voltage to 12 V if necessary (it should be set to 12 V from the previous procedure). The LED should be illuminated.

7. Record the DC supply voltage (E) in Table 2-5.

8. Using proper techniques, measure and record the circuit current (I), then return your meter to the voltage measurement mode.

9. Measure the remaining voltages in Table 2-5, then calculate the percent differences.

10. Leave the circuit connected for use in Procedure 2-4.

TABLE 2-5 Data for Figure 2-9

Parameter	Expected Value	Measured Value	Percent Difference
E			
I	21.3 mA		
V_R	10 V		
V_{LED}	2 V		
V_a	12 V		
V_b	2 V		

Observations

1. Did you use the lowest DMM scale that produced a reading? _____

2. Are the measured values close to your expected values? Quantify using percent error values.

_____

_____

3. Verify that $V_{ab} = V_a - V_b$ using measured values. _____

4. Verify that $V_b = V_a - V_{ab}$ using measured values. _____

Procedure 2-4

Series Circuit with *Reverse-Biased* LED

1. Remove the LED and replace it with the leads reversed, as shown in Figure 2-12. This reverse biases the LED, making it appear like an open (very high resistance).

2. Observe the LED. Is it illuminated?
   □ Yes   □ No

3. Re-measure voltages and currents and record them in Table 2-6.

4. Set the 12-V output voltage to zero and turn OFF the DC supply.

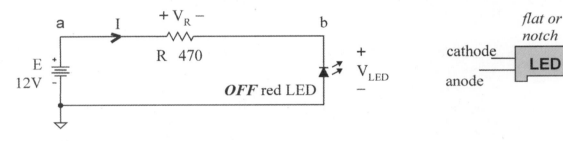

FIGURE 2-12   Series Circuit with Reverse-Biased LED

TABLE 2-6 Data Table for Figure 2-12			
Parameter	**Expected Value**	**Measured Value**	**Percent Difference**
E	12 V		
I	0 mA		
V_R	0 V		
V_{LED}	12 V		
V_a	12 V		
V_b	12 V		

Observations

1. Does the LED light with reverse bias?

_____

2. Does the reverse-biased LED act like an open, i.e., is measured $I = 0$?

_____

3. Is all the applied source voltage dropped across the open-acting LED?

_____

4. Does the resistor drop any voltage?

_____

Synthesis

Characteristic Curves of Resistor and LED

Build the circuit of Figure 2-13. Note that the LED is again forward biased, and the resistor is now 330 Ω (_orange_, _orange_, and _brown_). Measure the resistor and complete Table 2-7.

Determine the supply voltage required just to turn the LED on and record that at the bottom of Table 2-8 on page 28. Then, vary the supply voltage to the specified values and complete the remainder of Table 2-8.

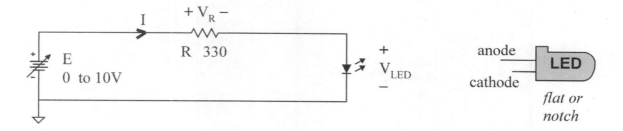

FIGURE 2-13   Series Circuit to Examine Device Characteristics

TABLE 2-7   Resistor Data Table for Figure 2-13

Resistor	Nominal Value	Measured Value	Percent Difference	Meets Spec? (within 5%)
R	330 Ω			☐ Yes ☐ No

TABLE 2-8 Characteristic Curve Data for Figure 2-13

Adjust P/S to set up E	Measured E	Measured V_R	Measured V_{LED}	Measured I
0 V	0 V	0 V	0 V	0 mA
1 V				
1.5 V				
2 V				
4 V				
7 V				
10 V		(STOP)		(STOP)
LED turn-on threshold				

(STOP) Instructor sign-off of the measurement for the 10-V supply _____

Using the data in Table 2-8, plot the data points for I_R vs. V_R on Figure 2-14(a), then plot the data points for I_{LED} vs. V_{LED} on Figure 2-14(b). Note that in a series circuit the current is the same everywhere ($I = I_R = I_{LED}$).

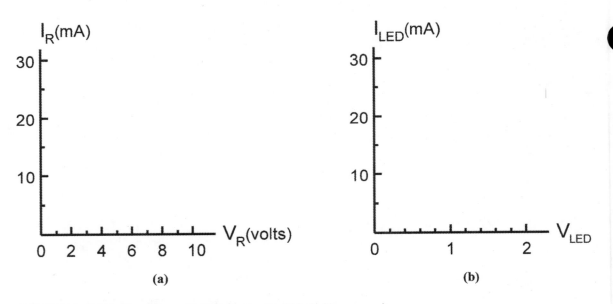

FIGURE 2-14 (a) Resistor Characteristic Curve for $R = 330 \ \Omega$ (linear curve)
(b) LED Characteristic Curve for Red LED (nonlinear curve)

Observations

1. Looking at the resistor characteristic curve of Figure 2-14(a), does it appear linear (does increasing V_R increase I_R by the same proportion)? ☐ Yes ☐ No

2. Using resistor characteristic curve of Figure 2-14(a), estimate I_R when $V_R = 6$ V.

   $I_R = $ _____

3. Using resistor characteristic curve of Figure 2-14(a), estimate V_R when $I_R = 10$ mA.

$V_R = \underline{\hspace{1.5cm}}$

4. Looking at the LED characteristic curve of Figure 2-14(b), does it appear linear (does increasing V_{LED} increase I_{LED} by the same proportion)?   ☐ Yes   ☐ No

5. Using the LED characteristic curve of Figure 2-14(b), estimate I_{LED} when $V_{LED} = 0.75$ V.

$I_{LED} = \underline{\hspace{1.5cm}}$

6. Using the LED characteristic curve of Figure 2-14(b), estimate V_{LED} when $I_{LED} = 10$ mA.

$V_{LED} = \underline{\hspace{1.5cm}}$

7. What are the LED current and voltage at the turn-on threshold?

I_{LED} Threshold = $\underline{\hspace{1.5cm}}$ V_{LED} Threshold = $\underline{\hspace{1.5cm}}$

Resistance, Switches, and Diodes

Name: _____   Date: _____

Lab Section: _____ _____   Lab Instructor: _____
             day   time

Text Reference

DC/AC Circuits and Electronics: Principles and Applications
Chapter 3: Resistance
Chapter 4: Resistance Applications

Materials Required

Triple power supply (2 @ 0–20 volts DC; 1 @ 5 volts DC)
1 each 6.8-Ω, 1-kΩ, 4.7-kΩ, 1-MΩ resistors
1 10-kΩ single-turn potentiometer
1 10-kΩ multi-turn potentiometer
1 Diode 1N4001 or equivalent
1 Single-pole double-throw (SPDT) switch
2 Light-emitting diodes (LED), one red, one green
1 Digital multimeter (preferably with a diode check option)
1 Protoboard 10 (pre-wired by student)
1 Safety glasses (for soldering)

Introduction

In this exercise, you will:

- Begin to identify resistors by the color code and measure resistors to determine if they meet specifications.
- Use an ohmmeter to examine variable resistors (potentiometers) and the forward- and reverse-bias resistance of a diode.
- Examine the characteristic (*V-I*) curves for a fixed resistor only and find that $V/I = R$ (Ohm's Law). In the previous exercise *V-I* characteristics for the LED and resistor were produced.
- Look at the function of a simple single-pole double-throw switch, then apply that to a circuit.

Pre-Lab Activity Checklist ◼

☐ Locate the resistors required in Figure 3-1.

☐ Calculate and find all "expected" table values (Tables 3-2 and 3-3). Include sample calculations supporting the values entered.

☐ Complete Figures 3-5 and 3-10 using bubble notation and the board layout for Figure 3-6.

☐ Solder 22-gauge wires to your single-pole double-throw switch. **Wear Safety Glasses!** (This can be performed in class if you lack soldering experience or if soldering tools are not available.)

☐ Pre-wire the circuit of Figure 3-9 shown in Figure 3-11 (trim resistors appropriately).

☐ Organize your resistors so that you can find them easily.

Performance Checklist ◼

🛑 Prelab completed? _____ Must be initialed by instructor before beginning lab work.

☐ Table 3-1: Demonstrate 1-MΩ resistance on 20-MΩ scale.

☐ Table 3-6: Demonstrate measured E, V_R, and I_R.

☐ Table 3-9: Demonstrate measured E, I_D, V_D, V_R, $V_{\text{LED RED}}$, $V_{\text{LED GREEN}}$.

Procedure 3-1

Resistor Color Code

1. Using the board wired prior to entering the laboratory (Figure 3-1), measure and record each resistance reading using the ohmmeter range in Table 3-1. For the 1-MΩ readings "with finger tips," touch both probes while making the measurement.

⤴ **Note:** A display showing "1." indicates that the meter is out of range, i.e., the resistor is larger than the chosen range.

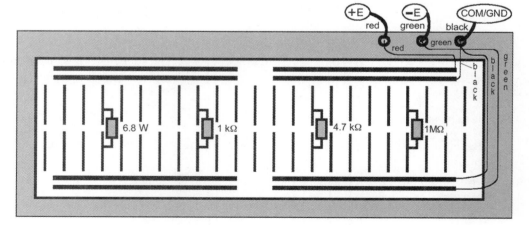

FIGURE 3-1 Resistor Layout for Procedure 3-1 (*pre-lab*)

2. Demonstrate to your instructor the 1-MΩ readings on the 20-MΩ scale with and without finger tips.

TABLE 3-I Carbon Resistor Measurements Using the DMM Ohmmeter

Resistance Nominal	MEASURED RESISTANCE					
	Ohmmeter Range					
	200 Ω	2 kΩ	20 kΩ	200 kΩ	2000 kΩ	20 MΩ
6.8 Ω						
I kΩ	I.					
4.7 kΩ	I.					
I MΩ	I.					STOP
I MΩ with finger tips	I.					STOP

STOP Instructor sign-off of I-MΩ readings _____

Observations ■

1. A "1." on the 20-MΩ ohmmeter scale indicates that (check the appropriate box):
   ☐ The circuit is open
   ☐ The resistance measured is greater than 20 MΩ
   ☐ The resistance measured is less than 20 MΩ

2. What does the "1." reading on the DMM ohmmeter setting indicate?

   _____

3. Which is the most accurate range to use for 2.2-kΩ resistance?
   ☐ 200 Ω   ☐ 2 kΩ   ☐ 20 kΩ   ☐ 200 kΩ   ☐ 2 MΩ   ☐ 20 MΩ

4. Describe how to select the most accurate range to use for a given resistance.

   _____

   _____

5. Compare the 1-MΩ reading with the 1-MΩ reading taken with your finger tips on the probe tips. List two reasons for keeping your fingers off the test probes when making circuit measurements.

   _____

   _____

Procedure 3-2

Resistor Tolerance

1. Enter the most accurate resistance reading from Table 3-1 in the "Measured Resistance" column of Table 3-2.
2. Calculate the percent difference of the measured value with respect to the nominal value. Include the 1-kΩ work in the sample calculations below.
3. Compare the measured resistance to the expected minimum and maximum values and choose the appropriate box in the "Meets Spec?" column.

TABLE 3-2 Carbon Resistor Measurements Compared with Nominal Resistance Ranges

Nominal Resistance	Expected (*Pre-Lab*)		Measured Resistance (from Table 3-1)	% Difference from Nominal Value	Meets Spec?	
	Minimum Resistance	Maximum Resistance				
6.8 Ω					☐ Yes	☐ No
1 kΩ					☐ Yes	☐ No
4.7 kΩ					☐ Yes	☐ No
1 MΩ					☐ Yes	☐ No

SAMPLE CALCULATIONS

Use the standard equation format to show your pre-lab calculations for the minimum and maximum expected resistances for the 1-kΩ resistor:

R_{Min} (1 kΩ) =

R_{Max} (1 kΩ) =

Show your calculation for percent difference for 1-kΩ resistor:

$$\% \text{ difference } (1 \text{ k}\Omega) = \frac{R_{\text{MEASURED}} - R_{\text{NOMINAL}}}{R_{\text{NOMINAL}}} \times 100\% =$$

Observations

1. Were all of your resistors within tolerance? ☐ Yes ☐ No
2. Which one had the largest error? _____
3. Does a resistor with the color code "*brown, black, orange, none*" that measures 7.9 kΩ meet specifications? Show your calculations below using standard notation. ☐ Yes ☐ No

Procedure 3-3

Single-Turn and Multi-Turn Potentiometers

Refer to Figure 3-2.

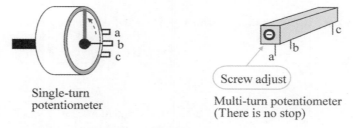

Single-turn
potentiometer

Multi-turn potentiometer
(There is no stop)

FIGURE 3-2 Single-Turn and Multi-Turn Potentiometers

1. Measure the resistance from the two outer terminals (a to c) on a 10-kΩ single-turn potentiometer. Record
   this value in Table 3-3. Turn the potentiometer knob and observe the effect on the reading.
2. Looking at the 10-kΩ single-turn potentiometer from the knob side, turn the potentiometer control to
   about half of the range. Measure the resistance between the left and center terminals (a to b) and observe
   the effect of turning the knob on the reading.
3. Turn the knob to adjust the resistance between a and b to 3 kΩ. Record that reading in Table 3-3.

TABLE 3-3 Potentiometer Data

Resistance Range	Expected Resistance	Single-Turn Measured Resistance	Multi-Turn Measured Resistance
R_{ac}	10 kΩ		
R_{ab}	3 kΩ		
R_{bc}			

4. Measure the resistance between terminals b and c, then record that reading in Table 3-3.

5. Measure the resistance from the two outer terminals (a to c) on a 10-kΩ multi-turn potentiometer. Record this value in Table 3-3. Turn the potentiometer knob and observe the effect on the reading.

6. Measure the resistance between the two closest terminals (a to b) and observe the effect of turning the knob on the reading.

7. Turn the knob to adjust the resistance between terminals a and b to 3 kΩ. Record that reading in Table 3-3.

8. Measure the resistance between terminals b and c and record that reading in Table 3-3.

Observations ◼

1. What effect did turning the knob have on the resistance between the outer terminals (a and c) of the single-turn potentiometer?

☐ It stayed fixed.　　☐ It changed.

2. What effect did turning the knob have on the resistance between the outer terminals of the multi-turn potentiometer?

☐ It stayed fixed.　　☐ It changed.

3. What effect did turning the knob have on the resistance measurement between terminals a and b of the single-turn potentiometer?

☐ It stayed fixed.　　☐ It changed.

4. What effect did turning the knob have on the resistance measurement between terminals a and b of the multi-turn potentiometer?

☐ It stayed fixed.　　☐ It changed.

5. As the resistance between a and b gets smaller, what do you think happens to the resistance between terminals b and c?

☐ It stays fixed.　　☐ It increases.　　☐ It decreases.

6. Which of the two potentiometers has a larger resistance change from a to b for a one-quarter turn of the knob?

☐ The single-turn potentiometer　　☐ The multi-turn potentiometer

7. If you wanted a 1.783-kΩ resistance, which potentiometer would be easier to adjust to that value?

☐ The single-turn potentiometer　　☐ The multi-turn potentiometer

8. If the a to b resistance of a 10-kΩ potentiometer is 1.783 kΩ, ideally what would the b to c resistance measure? _____

Show your work.

$R_{bc} =$ _____ = _____

Procedure 3-4

Diode Checking with an Ohmmeter

1. Locate a 1N4001 (or equivalent) diode. The end of the diode with the line is the cathode.

2. Label Figure 3-3 showing the cathode and anode terminals.

FIGURE 3-3 Diode Connections

3. Using the diode check on the multimeter (if available), connect the red and black leads together. Did the meter beep?　　　　　　　　　□ Yes　　□ No

4. Forward bias the diode by connecting the red lead to the anode and the black lead to the cathode. Did the meter beep?　　　　　　　　　□ Yes　　□ No

5. Reverse the leads and repeat the experiment using the diode check. Did the meter beep?

□ Yes　　□ No

6. Now use the ohmmeter on various scales to measure the forward-biased diode resistance (red lead to the anode; black lead to the cathode). Record the values in Table 3-4.

7. Now use the ohmmeter on various scales to measure the reverse-biased diode resistance (black lead to the anode; red lead to the cathode). Record the values in Table 3-4.

TABLE 3-4　Ohmmeter Diode Readings for Forward and Reversed Bias

	Ohmmeter Range					
	200 Ω	2 kΩ	20 kΩ	200 kΩ	2000 kΩ	20 MΩ
Forward Bias						
Reverse Bias						

Observations　　　　　　　　　　　　　　　　　　　　　　　　　　　■

1. Using the diode check on the multimeter (if available), which bias produced a beep?

_____

2. From Table 3-4, which ranges actually measured the diode's forward-bias resistance?
□ 200 Ω　　　□ 2 kΩ　　　□ 20 kΩ　　　□ 200 kΩ　　　□ 2 MΩ　　　□ 20 MΩ

3. Did the forward-bias resistance change based on which range you were on?　　□ Yes　　□ No

If so, why? _____

_____

4. What can you conclude about the resistance of a reverse-biased diode?

_____

_____

Procedure 3-5

Characteristic Curve of a Resistor

Refer to Figure 3-4. Be sure you have completed Figures 3-5 and 3-6 before proceeding.

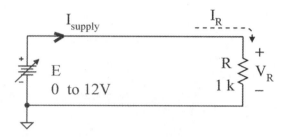

FIGURE 3-4   Circuit for Resistor Characteristic Curve

FIGURE 3-5   Circuit for Resistor Characteristic Curve Using Bubble Notation (*pre-lab*)

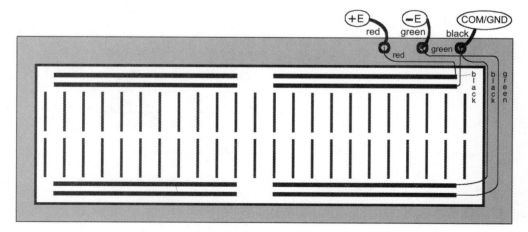

FIGURE 3-6   Intended Wiring Diagram for Resistor Characteristic Curve Circuit (*pre-lab*)

1. Using proper techniques, measure the value of the 1-kΩ resistor. Record that value and calculate the percent difference in Table 3-5.

TABLE 3-5 Resistor Data Table for Figure 3-4

Resistor	Nominal Value	Measured Value	Percent Difference	Meets Spec?
R	1 kΩ			☐ Yes ☐ No

2. Adjust a 0 to 20 VDC supply for 3 V and measure E, V_R, and I_R. Record those measurements in Table 3-6.
3. Repeat those measurements for 6 V and 12 V and record the values in Table 3-6.

TABLE 3-6 Characteristic Curve Data for 1-kΩ Resistor

Adjust P/S to set up E	Measured E	Measured V_R	Measured I_R
0 V	0 V	0 V	0 mA
3 V			
6 V			
12 V			
−3 V			
−6 V			
−12 V	STOP	STOP	STOP

4. Reverse the DC supply leads and measure and record the values for −3 V, −6 V, and −12 V. Demonstrate the −12 V measurement to your instructor.

STOP Instructor sign-off of −12 V measurements _____

5. Plot the data points for V_R and I_R from Table 3-6 on the axis of Figure 3-7. This is the characteristic curve for the 1-kΩ resistor.

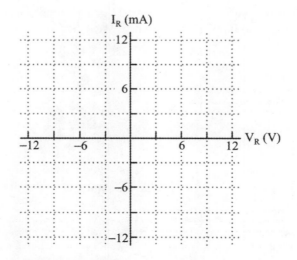

FIGURE 3-7 1-kΩ Characteristic Curve

Observations

1. Based upon your sketch of Figure 3-7, is the resistor a linear device?   ☐ Yes   ☐ No

   Why or why not? _____

   _____

2. Using the resistor characteristic curve, estimate I_R when $V_R = 9$ V.

   $I_R = $ _____

   _____

3. Using the resistor characteristic curve, find V_R when $I_R = -9$ mA.

   $V_R = $ _____

   _____

4. Using the measurements at 6 V and –6 V, check to see if the DC static resistance formula of $R = \dfrac{V_R}{I_R}$

   is satisfied? Use the standard format for your work below.   ☐ Yes   ☐ No

Procedure 3-6

SPDT Switch Check with Ohmmeter

1. If not completed during pre-lab, solder 22-gauge wires to terminals a, b, and c (see Figure 3-8) of your single-pole double-throw (SPDT) switch. **Wear Safety Glasses** and get help from your instructor if you have never soldered.

a b c

FIGURE 3-8   Switch in Toggle Left Position

2. Place the switch in the toggle-left position as shown in Figure 3-8. Measure the resistance of the left and right terminal pairs to determine if they are open or closed. Check the appropriate box in Table 3-7.

3. Repeat the measurements for the toggle-center and toggle-right positions.

TABLE 3-7 SPDT Switch Connectivity

Switch Position	Left Terminal Pair (a-b)		Right Terminal Pair (b-c)	
Toggle left	☐ Open	☐ Closed	☐ Open	☐ Closed
Toggle center	☐ Open	☐ Closed	☐ Open	☐ Closed
Toggle right	☐ Open	☐ Closed	☐ Open	☐ Closed

Observations ▪

1. If the SPDT switch is toggled to the left, which pair of terminals are connected?

   ☐ Right pair   ☐ Left pair   ☐ Neither pair

2. If the SPDT switch is toggled to the center, which pair of terminals are connected?

   ☐ Right pair   ☐ Left pair   ☐ Neither pair

3. If the SPDT switch is toggled to the right, which pair of terminals are connected?

   ☐ Right pair   ☐ Left pair   ☐ Neither pair

Synthesis
Switch, Diode, and LED Circuit

Refer to Figure 3-9. Be sure you have completed Figures 3-10 and 3-11 before proceeding.

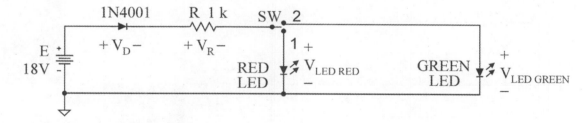

FIGURE 3-9 Diode, Switch, and LED Circuit

FIGURE 3-IO Diode, Switch, and LED Schematic Using Bubble Notation for Supplies (*pre lab*)

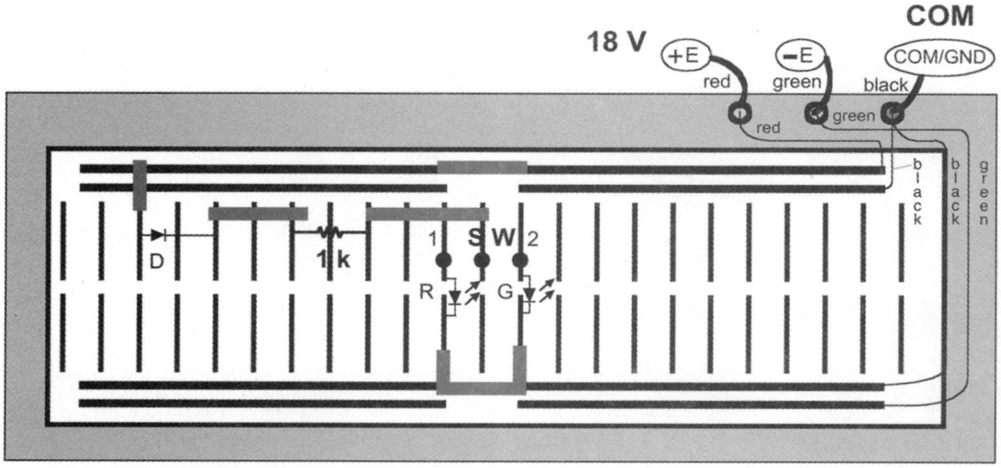

FIGURE 3-11   Diode, Switch, and LED Board Layout

Build the circuit of Figure 3-11 and have your instructor inspect your wired board before connecting the power. Failure to do so may result in damaged components! Set the DC supply voltage to 18 V, then connect it to the circuit with the switch in the center position.

Is there any current (I_D) with the switch in the center position? (If you are not sure, measure it.)

☐ Yes   ☐ No

Place the switch in position 1 and measure, then complete Table 3-8.

TABLE 3-8   Switch Position 1: Forward-Biased Diode

Variable	Expected Value	Measured Value	Percent Difference	ON or OFF
E	18 V			
I_D	15.3 mA			
V_D	0.7 V			
V_R	15.3 V			
$V_{LED\ RED}$	2 V			☐ On ☐ Off
$V_{LED\ GREEN}$	0 V			☐ On ☐ Off

Place the switch in position 2 and measure, then complete Table 3-9. Have your instructor sign off on these measurements.

TABLE 3-9 Switch Position 2: Forward-Biased Diode

Variable	Expected Value	Measured Value	Percent Difference	LED ON or OFF
E	18 V	(STOP)		
I_D	15 mA	(STOP)		
V_D	0.7 V	(STOP)		
V_R	15.0 V	(STOP)		
$V_{LED\ RED}$	0 V	(STOP)		☐ On ☐ Off
$V_{LED\ GREEN}$	2.3 V	(STOP)		☐ On ☐ Off

(STOP) Instructor sign-off of Table 3-9 _____

Reverse the switching diode (1N4001) to reverse bias the diode. With the switch in position 2, measure and record the values in Table 3-10.

TABLE 3-10 Switch Position 2: Reverse-Biased Diode

Variable	Expected Value	Measured Value	ON or OFF
E	18 V		
I_D	0 mA		
V_D	18 V		
V_R	0 V		
$V_{LED\ RED}$	0 V		☐ On ☐ Off
$V_{LED\ GREEN}$	0 V		☐ On ☐ Off

Observations

1. When the switch was in the center position, did the LEDs light?   ☐ Yes   ☐ No
2. Why? _____
3. In switch position 1, which of the LEDs were lit?   ☐ Red   ☐ Green   ☐ None
4. In switch position 1, did the diode D act like a short or an open?   ☐ Short   ☐ Open
5. In switch position 2, which of the LEDs were lit?   ☐ Red   ☐ Green   ☐ None
6. In switch position 2, did the diode D act like a short or an open?   ☐ Short   ☐ Open
7. When the diode was reverse biased, did the LEDs light?   ☐ Yes   ☐ No
8. Why? _____
9. When the diode was reverse biased, did it act like a short or an open?   ☐ Short   ☐ Open

Kirchhoff's Current Law

Name: _____ Date: _____

Lab Section: _____ _____ Lab Instructor: _____
　　　　　　　　day　　　　　　time

Text Reference ■

DC/AC Circuits and Electronics: Principles and Applications
Chapter 5: KCL—Kirchhoff's Current Law

Materials Required ■

Triple power supply (2 @ 0–20 volts DC; 1 @ 5 volts DC)
4　1-kΩ resistors
1　1.8-kΩ resistor
2　3.9-kΩ resistors
1　Light-emitting diode (LED), red
1　LM324 operational amplifier

Introduction ■

Kirchhoff's Current Law (KCL) describes how currents must behave at a circuit node. KCL states that the sum of the current into a node must equal the currents leaving that node. Mathematically,

$$\sum I_{\text{into node}} = \sum I_{\text{out of node}}$$

or alternately as

$$\sum I_{\text{into node}} - \sum I_{\text{out of node}} = 0 \text{ A}$$

In this exercise, you will:

- Apply KCL to a parallel resistive network
- Apply KCL to a simple node to find connecting branch currents
- Apply KCL to an inverting summing amplifier circuit
- Apply KCL to a bridge circuit.

Pre-Lab Activity Checklist

☐ Redraw the schematic of Figure 4-1 using bubble notation in Figure 4-2.

☐ Sketch the intended wiring diagram of Figure 4-1 in Figure 4-3.

☐ Find the expected values for Table 4-2; include sample calculations.

☐ Find the expected values for Table 4-3; include sample calculations.

☐ Build the circuit of Figure 4-5 shown in Figures 4-7 and 4-8.

☐ Determine the expected value of I_{Rf} in the Table 4-5 (show sample calculation).

Performance Checklist

☐ Pre-lab completed?   🛑 Instructor sign-off _____

☐ Table 4-2: Demonstrate measured V_a, V_b, E, and $I_{supply \ with \ R1}$.

☐ Table 4-5: Demonstrate measured I_{R1}, I_{R2}, and I_{Rf}.

☐ Table 4-6: Demonstrate measured I_{wire}.

Procedure 4-1

Parallel Resistive Circuit

In this procedure you will apply Kirchhoff's Current Law (KCL) to a parallel resistor network.

1. Measure each 1-kΩ resistor of Figure 4-1 and record those values in Table 4-1.

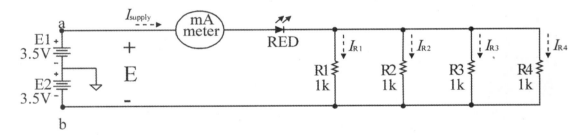

FIGURE 4-1   Parallel Resistive Branches

TABLE 4-1 Resistor Verification for Figure 4-1				
Resistor	**Nominal Value**	**Measured Value**	**Percent Difference**	**Meets Spec?**
R_1	1 kΩ			☐ Yes ☐ No
R_2	1 kΩ			☐ Yes ☐ No
R_3	1 kΩ			☐ Yes ☐ No
R_4	1 kΩ			☐ Yes ☐ No

2. Build the circuit of Figure 4-1. Initially R_2, R_3, and R_4 will not be connected.
3. Measure and record each of the parameters in Table 4-2. Add resistors R_2, R_3, and R_4 as required.

(STOP) Obtain instructor sign-off. _____

TABLE 4-2 Data Table for Figure 4-1 (‖ is a symbol for "parallel to")

Variable	Expected Value	Measured Value	Percent Difference	LED Brightness
E_1				
E_2				
V_a		(STOP)		
V_b		(STOP)		
E		(STOP)		
I_{supply} with R_1 only (see note below)	5 mA	(STOP)		
I_{supply} with R_1 and R_2 ($R_1 \parallel R_2$)	10 mA			
I_{supply} with R_1, R_2, and R_3 ($R_1 \parallel R_2 \parallel R_3$)				
I_{supply} with $R_1 \parallel R_2 \parallel R_3 \parallel R_4$				

Note: "I_{supply} with R_1" means that R_2, R_3, and R_4 have been removed from the circuit. Each resistive branch is expected to draw about 5 mA. When connected: $I_{R1} = I_{R2} = I_{R3} = I_{R4} \approx 5$ mA

4. Compare the brightness of the LED with one resistor to the brightness when using all four resistors.

Complete Figures 4-2 and 4-3 before proceeding.

(+3.5V)

(COM)

(–3.5V)

FIGURE 4-2 Bubble Notation Drawing of the Schematic of Figure 4-1 (*pre-lab*)

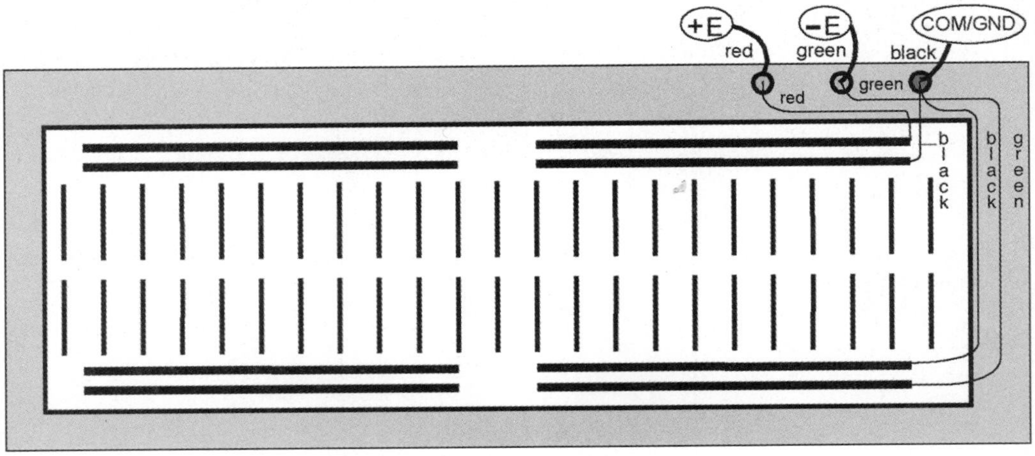

FIGURE 4-3 Component Layout of the Schematic of Figure 4-1 (*pre-lab*)

🛑 Instructor sign-off of measured values in Table 4-2 _____

SAMPLE CALCULATIONS

Show your sample pre-lab calculations for:

$E =$

I_{supply} with $R_1 \parallel R_2 \parallel R_3 \parallel R_4 =$

Observations ■

1. As each resistive branch current was added, what was the effect on the supply current? Quantify your answer. If it changed, how much?

   _____

   _____

2. Compare the LED brightness with one resistor to the brightness when using four resistors.

   _____

   _____

Procedure 4-2

Parallel Resistive Circuit (*continued*)

In this section you will apply KCL to a simple node to find connecting branch currents. This procedure uses the same circuit as Procedure 1.

1. With resistors R_1, R_2, R_3, and R_4 all connected as shown in Figure 4-4, measure and record each of the parameters in Table 4-3.

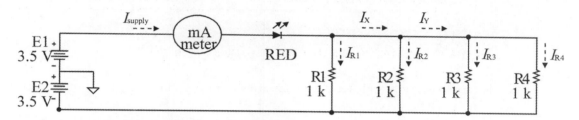

FIGURE 4-4 Same Circuit as Figure 4-1 with All Four Branches Connected

TABLE 4-3 Data Table for Figure 4-4

Variable	Expected Value	Measured Value	Percent Difference	LED Brightness
I_{R1}				
I_{R2}				The LED brightness should not change while making these measurements.
I_{R3}				
I_{R4}				
I_X	15 mA			
I_Y				

SAMPLE CALCULATIONS

Show your sample pre-lab calculations for:

$I_X =$

$I_Y =$

Observations

1. Using measured values, verify KCL at the bottom *voltage node*.

2. If the LED were reversed in this circuit, what would the supply current be?

Procedure 4-3

Inverting Summing Amplifier

In this section you will apply KCL to an inverting summing amplifier circuit using an LM324 operational amplifier (op amp).

WARNING!   The LM 324 must be handled with care.

The chip must be powered at pins #4 and #11 prior to OR at the same time as the application of a signal to input pins #12 or #13. If a signal is applied to pins #12 or #13 prior to connection of the supply voltages at pins #4 and #11, the chip will be permanently damaged.

The input voltages to the op amp must not be greater than $+E_{supply}$ or less than $-E_{supply}$.

1. Measure each resistor of Figure 4-5 and record those values in Table 4-4.

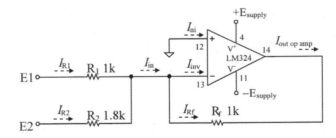

FIGURE 4-5   Inverting Summing Amplifier (*pre-lab*)

TABLE 4-4   Resistor Data Table for Figure 4-5

Resistor	Nominal Value	Measured Value	Percent Difference	Meets Spec?
R_1	1 kΩ			☐ Yes ☐ No
R_2	1.8 kΩ			☐ Yes ☐ No
R_f	1 kΩ			☐ Yes ☐ No

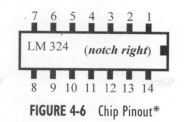

FIGURE 4-6 Chip Pinout*

➔ Note: For IC pin numbering, begin at the notched or dotted end and begin counting counterclockwise starting with pin #1 as shown in Figure 4-6.

Refer to Figures 4-7 and 4-8 when building the circuit in Figure 4-5.

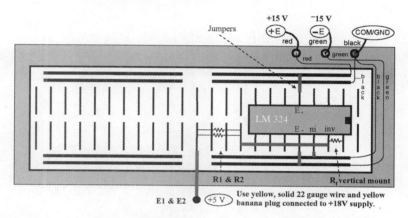

FIGURE 4-7 Wiring Board Layout for the Inverting Summing Amplifier Circuit of Figure 4-8 (not to scale)

2. Connect the voltage supplies to the circuit of Figure 4-5 (built in pre-lab).
3. Measure and record the parameters in Table 4-5.

TABLE 4-5 Data Table for Figure 4-5

Variable	Expected Value	Measured Value	Percent Difference
$+E_{supply}$	15 V		
$-E_{supply}$	-15 V		
E_1	5 V		
E_2	5 V		
I_{ni}	0 mA		
I_{inv}	0 mA		
I_{R1}	5 mA	STOP	
I_{R2}	2.78 mA	STOP	
I_{Rf}		STOP	

🛑 Instructor sign-off of measured values in Table 4-5 _____

SAMPLE CALCULATIONS

Show your sample calculation for I_{Rf} using KCL.

$I_{Rf} =$

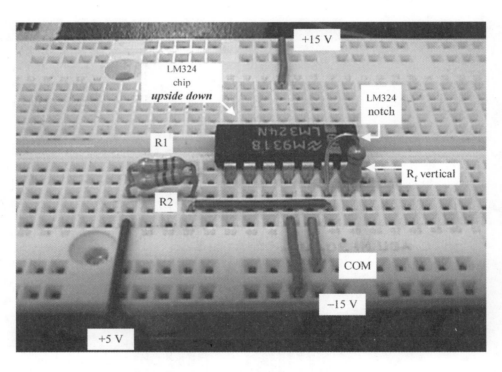

FIGURE 4-8 Photograph of Inverting Summing Amplifier Circuit of Figure 4-7

Observations ■

1. Does the summing circuit add the input currents (I_1 and I_2) in the R_f branch?

   _____

2. What is the value of $I_{\text{out op amp}}$?

   _____

Synthesis
Bridge Circuit

First, apply KCL to the circuit of Figure 4-9 to find the expected values in Table 4-6. Then, build the circuit, verify the values, and record them in Table 4-6.

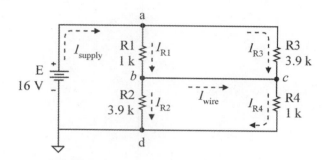

FIGURE 4-9 Bridge Circuit

TABLE 4-6 Data Table for Circuit of Figure 4-5

Variable	Expected Value	Measured Value	Percent Difference
E	16 V		
I_{supply}	10 mA		
I_{R1}	8 mA		
I_{R2}	2 mA		
I_{R3}			
I_{R4}			
I_{wire}		🛑	

🛑 Demonstrate I_{wire} to your instructor _____

Observations

1. Verify KCL at node **a** with measured values.

   $I_{supply} =$ _____ $\approx I_{R1} + I_{R3} =$ _____ (values should be equal +/– 50 µA)

2. Verify KCL at node **b** with measured values (use the format above).

3. Verify KCL at node **c** with measured values.

_____

4. Verify KCL at node **d** with measured values.

_____

Kirchhoff's Voltage Law

Name: _____   Date: _____

Lab Section: _____ _____   Lab Instructor: _____
　　　　　　　　　 day　　　　　　　 time

Text Reference ◼

DC/AC Circuits and Electronics: Principles and Applications
Chapter 6: KVL—Kirchhoff's Voltage Law

Materials Required ◼

Triple power supply (2 @ 0–20 volts DC; 1 @ 5 volts DC)

3　1-kΩ resistors

2　3.9-kΩ resistors

1　each 220-Ω, 330-Ω, 2.2-kΩ, and 3.3-kΩ resistor

1　Light-emitting diode (LED), red

1　Switch (a wire may be used)

1　LM324 operational amplifier

Introduction ◼

Kirchhoff's Voltage Law (KVL) describes how voltages must behave in a closed-loop circuit. KVL states that the sum of the voltage rises minus the number of voltage falls around a closed circuit loop must equal zero. Mathematically:

$$\sum V_{rises} - \sum V_{falls} = 0 \text{ V}$$

or

$$\sum V_{rises} = \sum V_{falls}$$

By measuring both voltage drops and node voltages, you will:

- Apply KVL to a series circuit with both a single and split supply voltage
- Apply KVL to a bridge circuit
- Apply KVL to a non-inverting voltage amplifier circuit.

Pre-Lab Activity Checklist

☐ Redraw the schematic using bubble notation.
  ☐ Figure 5-1 in Figure 5-2
  ☐ Figure 5-4 in Figure 5-5
☐ Draw the intended protoboard layout in Figure 5-3.
☐ Find the expected values in:
  ☐ Table 5-2   ☐ Table 5-7
  ☐ Table 5-3   ☐ Table 5-8
  ☐ Table 5-4   ☐ Table 5-10
  ☐ Table 5-5
☐ Build the circuits of:
  ☐ Figure 5-1
  ☐ Figure 5-6

Performance Checklist

☐ Pre-lab completed?   🛑 Instructor sign-off _____
☐ Table 5-3: Demonstrate measured V_d, V_{bd}, and V_{db}.
☐ Table 5-8: Demonstrate measured V_{in}, V_a, V_{Rf}, and V_{out}.
☐ Table 5-10: Demonstrate measured V_{load}.

Procedure 5-1

Series Circuit Voltage Drops: Switch Closed

Refer to Figure 5-1. Be sure you have completed Figures 5-2 and 5-3 before proceeding.

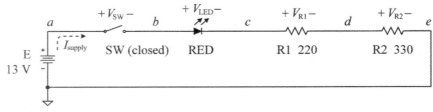

FIGURE 5-1   Series Circuit Used in Procedures 5-1, 5-2, and 5-3

FIGURE 5-2   Series Circuit Using Bubble Notation (*pre-lab*)

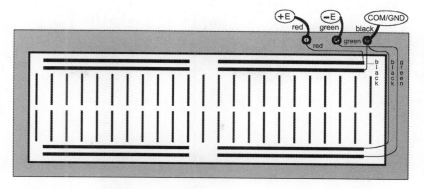

FIGURE 5-3 Intended Wiring Diagram for Series Circuit of Figure 5-1 (*pre-lab*)

1. Measure the resistance values in Table 5-1 and calculate the percent differences.
2. Set up and connect the power supply.
3. Measure the voltage drops across the components in Figure 5-1 and record them in Table 5-2. Use the polarity shown on the schematic.
4. Calculate the percent differences in Table 5-2.

TABLE 5-1 Resistor Data for Figure 5-1

Resistor	Nominal Value	Measured Value	Percent Difference	Meet Specs?
R_1	220 Ω			☐ Yes ☐ No
R_2	330 Ω			☐ Yes ☐ No

TABLE 5-2 Series Circuit Component Voltages with Switch Closed

Parameter	Expected Value	Measured Value	Percent Difference
E			
V_{SW}			
V_{LED}			
V_{R1}	4.4 V		
V_{R2}			

SAMPLE CALCULATIONS

Using KVL, find the voltage drop across V_{R2}.

$V_{R2} =$

% Difference $R_2 =$

Observations ▪

1. Use measured voltages and verify KVL by $\Sigma V_{rises} - \Sigma V_{falls} = 0$ V around the closed loop.

    ⊐ Note: Due to instrument tolerances the sum may not be exactly 0 V; +/− 0.1 V is acceptable.

---

2. Use measured voltages and verify the alternate KVL by $\Sigma V_{rises} = \Sigma V_{falls}$ around a closed loop.

Procedure 5-2

Series Node Voltages: Switch Closed

1. Using measured voltages from Table 5-2, start at common and "walk" clockwise around the circuit until you return to common. Record node voltages in Table 5-3 as you walk.

TABLE 5-3 Series Circuit Node Voltages with SWl Closed

Parameter	Expected Value	Measured Value	Percent Difference	Clockwise "Walk" Values	Counterclockwise "Walk" Values
V_a					
V_b					
V_c					
V_d		STOP			
V_e					
V_{cd}					
V_{bd}		STOP			
V_{db}		STOP			

2. Repeat the "walk" in a counterclockwise direction.

3. Measure the node voltages and node voltage differences for the circuit in Figure 5-1 and record them in Table 5-3. Demonstrate the voltage measurements for V_d, V_{bd}, and V_{db} to your instructor.

4. Calculate the percent differences in Table 5-3.

🛑 Instructor sign-off of measured values in Table 5-3 _____

SAMPLE CALCULATIONS

Using the expected voltages from Table 5-2, find:

V_c by starting at common and "walking" clockwise to node c:

$V_c =$

V_{bd} by starting at node d and walking counterclockwise to node b:

$V_{bd} =$

Find V_{db} by subtracting node voltages:

$V_{db} = V_d - V_b =$

Observations ◾

1. Were the expected node voltages found by "walking" the circuit equal to the measured values?

   ⌦ **Note:** These values may not be exactly equal due to measurement tolerances; differences of +/–0.1 V are acceptable.

   ☐ Yes ☐ No

Procedure 5-3

Series Circuit Voltages: Switch Open

1. Open the switch in the series circuit of Figure 5-1.

2. Measure the node voltages and node voltage differences for the circuit in Figure 5-1 and record them in Table 5-4.

3. Calculate the percent differences in Table 5-4.

TABLE 5-4 Series Circuit Node Voltages with SW1 Open

Parameter	Expected Value	Measured Value	Percent Difference
E			
V_{SW}		◆	
V_{LED}	0 V		
V_{R1}	0 V		
V_{R2}	0 V		
V_a			
V_b			
V_c			
V_d			
V_e			
V_{bd}			

◆ The DMM DC voltmeter typically has a resistance of about 10 MΩ and completes the circuit. This makes the voltage reading here substantially different than the expected value.

SAMPLE CALCULATIONS

Using the expected component voltages in Table 5-4, find:

V_{SW} using KVL:

$V_{SW} =$

V_a by starting at common and walking clockwise to node a:

$V_a =$

V_b by starting at common and walking counterclockwise to node b:

$V_b =$

Observations

1. Ideally, which component dropped the entire supply voltage (E) while the other components dropped 0 V?

   _____

   When measured, why did this not occur?

   _____

2. Suggest a troubleshooting technique to find an open using node voltages as test voltages. *Hint*: Look at the node voltages on both sides of the open.

   _____

Procedure 5-4

Series Circuit with Split Supply: Switch Closed

1. Disconnect the power supply and adjust two separate supply voltages to 6.5 V each.
2. Connect the supplies to your previously built circuit as shown in Figure 5-4. Note that the only change is in the configuration of the supply voltage.
3. Close the switch and measure the node voltages and node voltage differences for the circuit and record them in Table 5-5. Demonstrate the voltage measurements for V_d, V_{bd}, and V_{db} to your instructor.
4. Calculate the percent differences in Table 5-5.

Complete Figure 5-5 before proceeding.

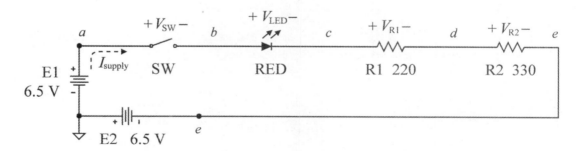

FIGURE 5-4 Series Circuit with Split-Supply Switch Closed

FIGURE 5-5 Split Supply Schematic of Figure 5-4 Using Bubble Notation (*pre-lab*)

TABLE 5-5 Series Circuit with Split Supply

Parameter	Expected Value	Measured Value	Percent Difference
E_1			
E_2			
$V_{SW}{}^*$			
$V_{LED}{}^*$			
$V_{R1}{}^*$			
$V_{R2}{}^*$			
V_a			
V_b			
V_c			
V_d			
V_e			
$E = V_{ae}$			
V_{cd}			
V_{bd}			
V_{db}			

*These should be the same expected component voltage drops as Table 5-1.

SAMPLE CALCULATIONS

Using KVL walks, find the expected values for:

$V_a =$

$V_b =$

Net supply $E = V_{ae} =$

Observations

1. Based on your measurements, did replacing E of 13 V with a split supply of 6.5 V and − 6.5 change the voltage drops across the individual devices? ☐ Yes ☐ No

2. Did the node voltage values change? ☐ Yes ☐ No

3. Did any voltage differences change, for example, V_{bd}? ☐ Yes ☐ No

Procedure 5-5

Noninverting Voltage Amplifier

In this section you will apply KCL to an inverting summing amplifier circuit using an LM324 operational amplifier (op-amp).

> ⚠ **WARNING! The LM 324 must be handled with care.**
>
> The chip must be powered at pins #4 and #11 prior to OR at the same time as the application of a signal to input pins #12 or #13. If a signal is applied to pins #12 or #13 prior to connection of the supply voltages at pins #4 and #11, the chip will be permanently damaged.
>
> The input voltages to the op amp must not be greater than $+E_{supply}$ or less than $-E_{supply}$.

1. Measure each resistor of Figure 5-6 and record those values in Table 5-6. Notice the LM324 pinout shown in Figure 5-7.

2. Connect the voltage supplies to the circuit of Figure 5-6 (*built in pre-lab*). If only two variable supplies are available, use a potentiometer as a voltage divider to provide V_{in}.

3. With V_{in} at 1 V, measure and record each of the parameters in Table 5-7. Calculate percent differences.

4. Set V_{in} to 2 V and measure and record each of the parameters in Table 5-8. Calculate percent differences. (*instructor sign-off required*).

⊐ **Note:** Save this circuit for use in Exercise 6 (supply voltage and resistor change only).

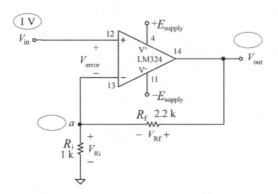

FIGURE 5-6 Noninverting Voltage Amplifier Circuit (*pre-lab*)

FIGURE 5-7 LM324 Pinout

TABLE 5-6 Resistor Data for Circuit of Figure 5-6

Resistor	Nominal Value	Measured Value	Percent Difference	Meets Spec?
R_i	1 kΩ			☐ Yes ☐ No
R_f	2.2 kΩ			☐ Yes ☐ No

Refer to the inverting amplifier circuits shown in Figures 5-8 and 5-9.

TABLE 5-7 Data for Circuit of Figure 5-6 with 1-V Input

Variable	Expected Value	Measured Value	Percent Difference
$+E_{supply}$	15 V		
$-E_{supply}$	−15 V		
V_{in}	1 V		
V_{error}	≅ 0 V	◆	
V_a			
V_{Rf}	2.2 V		
V_{out}			

◆ Use lowest possible range to make measurement.

TABLE 5-8 Data for Circuit of Figure 5-6 with 2-V input

Variable	Expected Value	Measured Value	Percent Difference
$+E_{supply}$	15 V		
$-E_{supply}$	− 15 V		
V_{in}	2 V	🛑	
V_{error}	≅ 0 V	◆ 🛑	
V_a		🛑	
V_{Rf}	4.4 V	🛑	
V_{out}		🛑	

◆ Use lowest possible range to make measurement.

🛑 Instructor sign-off of measured values in Table 5-8 _____

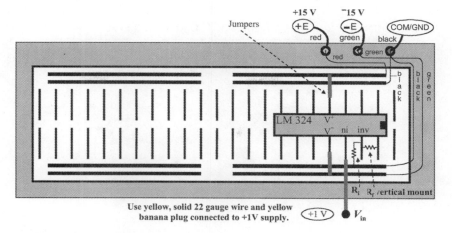

FIGURE 5-8 Wiring Board Layout for the Inverting Amplifier Circuit of Figure 5-6 (*pre-lab*)

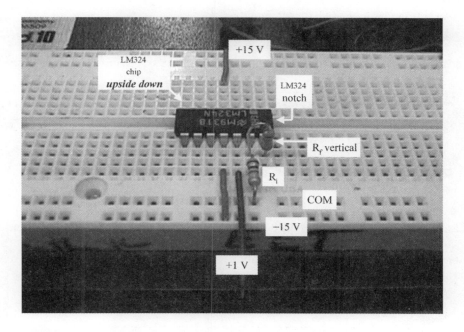

FIGURE 5-9 Photograph of Inverting Summing Amplifier Circuit of Figure 5-6

SAMPLE CALCULATIONS

Using the given expected values from Table 5-7, find:

V_a (with V_{in} at 1 V) $= V_{in} - V_{error} =$

V_{out} (with V_{in} at 1 V) $= V_a + V_{Rf} =$

SAMPLE CALCULATIONS

Using the given expected values from Table 5-8, find:

V_a (with V_{in} at 2 V) =

V_{out} (with V_{in} at 2 V) =

Observations ■

1. Was the input voltage amplified at V_{out}? (That is, was V_{out} larger than V_{in}?) ☐ Yes ☐ No
2. What is the voltage gain (V_{out}/V_{in})?

---

3. Was the output voltage inverted (negative) or noninverted (positive)? ☐ Inverted ☐ Noninverted

Synthesis
Bridge Circuit

Measure all the resistors used in the circuit of Figure 5-10 and complete Table 5-9. Draw the bubble notation version of the circuit on Figure 5-11. Build the circuit of Figure 5-10; set and connect the supply. Measure the voltages for the circuit and complete Table 5-10. Demonstrate the voltage measurement for V_{load} to your instructor.

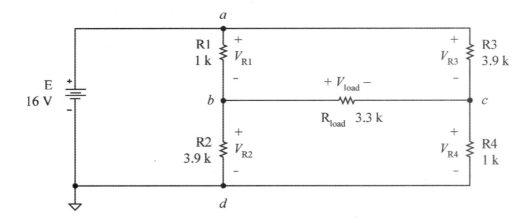

FIGURE 5-10 Bridge Circuit

FIGURE 5-11 Bridge Circuit of Figure 5-10 Using Bubble Notation

TABLE 5-9 Resistor Data for Circuit of Figure 5-10

Resistor	Nominal Value	Measured Value	Percent Difference	Meets Spec?
R_1	1 kΩ			☐ Yes ☐ No
R_2	3.9 kΩ			☐ Yes ☐ No
R_3	3.9 kΩ			☐ Yes ☐ No
R_4	1 kΩ			☐ Yes ☐ No
R_{load}	3.3 kΩ			☐ Yes ☐ No

TABLE 5-10 Data for Bridge Circuit of Figure 5-10

Variable	Expected Value	Measured Value	Percent Difference
E	16.0 V		
V_{R1}	4.8 V		
V_{R2}			
V_{R3}	11.2 V		
V_{R4}			
V_{load}		🛑	

🛑 Instructor sign-off of measured V_{load} _____

Observations ◼

1. Verify KVL of the left window (E, R_1, and R_2) with measured values.

   _____

2. Verify KVL of the top right window (R_1, R_3, and R_{load}) with measured values.

   _____

3. Verify KVL of the bottom right window (R_2, R_4, and R_{load}) using measured values.

   _____

4. Walk from $-V_{load}$ terminal to the $+V_{load}$ terminal and verify the V_{load}.

   _____

Ohm's Law and the Power Rule

Name: _____ Date: _____

Lab Section: _____ _____ Lab Instructor: _____
                  day   time

Text Reference

DC/AC Circuits and Electronics: Principles and Applications
Chapter 7: Ohm's Law, Power, and Energy

Materials Required

Triple power supply (2 @ 0–20 volts DC; 1 @ 5 volts DC)
1   220-Ω, 1-kΩ, 1.2-kΩ, 2.2-kΩ resistor
1   Light-emitting diode (LED), red
1   LM324 operational amplifier

Introduction

Ohm's Law describes the relationship between current, voltage, and resistance in a circuit. Mathematically,

$V = I \cdot R$

In this exercise, you will:

- Apply Ohm's Law to a resistor using both a fixed and variable supply voltage
- Apply Ohm's Law to a series-parallel circuit with an LED
- Apply Ohm's Law to a noninverting voltage amplifier circuit.

Pre-Lab Activity Checklist ◼

☐ Redraw the schematic of Figure 6-2 in Figure 6-3 using bubble notation.

☐ Draw the intended wiring diagrams for Figure 6-2 in Figure 6-4.

☐ Find the expected values in:

  ☐ Table 6-2

  ☐ Table 6-4

  ☐ Table 6-6

  ☐ Table 6-8

  ☐ Table 6-9

☐ Build the circuits of:

  ☐ Figure 6-2

  ☐ Figure 6-5 (same as Figure 5-7 except $R_f = 1.2$ kΩ)

Performance Checklist ◼

☐ Pre-lab completed?   🛑 Instructor sign-off _____

☐ Table 6-2:   Demonstrate measured current I for 10-V source.

☐ Table 6-6:   Demonstrate measured V_{R2} and I_{R1}.

☐ Table 6-8:   Demonstrate measured V_{Ri}, V_{Rf}, and V_{out}.

☐ Table 6-9:   Demonstrate measured V_{Ri}, V_{Rf}, and V_{out}.

Procedure 6-1

Predicting Circuit Current Using Ohm's Law

1. Measure the 1-kΩ resistor used in Figure 6-1 and record the value in Table 6-1. Calculate the percent difference.

2. Set the supply voltage to 1 V, then connect it to the circuit.

3. Using the polarity and current direction shown on Figure 6-1, measure the voltages and currents specified in Table 6-2.

4. Demonstrate the circuit current with 10 V applied to your instructor, then calculate the percent differences in I_R.

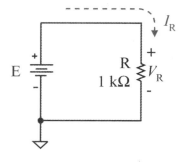

FIGURE 6-1   Ohm's Law Test Circuit

TABLE 6-1 Resistor Data Table for Figure 6-1

Resistor	Nominal Value	Measured Value	Percent Difference	Meets Spec?
R	1 kΩ			☐ Yes ☐ No

TABLE 6-2 Ohm's Law to Find I (Data Table for Figure 6-1)

Adjust E to Set Up V_R	Expected I_R	Measured V_R	Measured I_R	I_R Percent Difference
1 V				
5 V				
10 V				

🛑 Instructor sign-off of current with 10 V applied _____

SAMPLE CALCULATIONS

Show your expected pre-lab calculations for I_R when $V_R = 1$ V.

I_R(at 1V) =

Observation

1. Did Ohm's Law accurately predict the resistor current?   ☐ Yes   ☐ No

Procedure 6-2

Predicting Power Dissipation

1. Replace the 1-kΩ resistor of Figure 6-1 with a 220-Ω resistor. Measure the 220-Ω resistor, record the value in Table 6-3, and calculate the percent difference.
2. Using the supply voltages specified, measure the voltages and currents in Table 6-4. Use the polarity and current direction shown on Figure 6-1.
3. Calculate the dissipated power in the resistor and record those values in Table 6-4.

TABLE 6-3 Resistor Data Table for Figure 6-1 with a 220-Ω Resistor

Resistor	Nominal Value	Measured Value	Percent Difference	Meets Spec?
R	220 Ω			☐ Yes ☐ No

TABLE 6-4 Data Table for Figure 6-1 with a 220-Ω Resistor

Adjust E to Set Up V_R	Expected I_R	Expected P	Measured V_R	Measured I_R	Measured P (Use Measured V_R and I_R)	Resistor Heat Dissipated BE CAREFUL
1 V						☐ Cool ☐ Warm ☐ Hot
5 V						☐ Cool ☐ Warm ☐ Hot
8 V						☐ Cool ☐ Warm ☐ Hot
10 V						☐ Cool ☐ Warm ☐ Hot

SAMPLE CALCULATIONS

Show your pre-lab expected calculations for I_R when V_R = 1 V.

I_R (at 1 V) =

P (at 1 V) =

Observations ◼

1. As the power dissipation increased in the resistor, did the resistor get warmer?

   ☐ Yes ☐ No

2. What is the maximum power that is safely dissipated by this resistor?

   ☐ 1/8 W ☐ 1/4 W ☐ 1/2 W ☐ 1 W ☐ _____W

3. Using the maximum power for this resistor and the power rule ($P = I^2R$), find the maximum current that this resistor can safely handle. Show your calculations.

   _____

4. Using the maximum power for this resistor and the power rule ($P = V^2/R$), find the maximum voltage that this resistor can safely handle. Show your calculations.

   _____

Procedure 6-3

Ohm's Law, KCL, and KVL

1. For the circuit of Figure 6-2, measure the resistors and record the values in Table 6-5. Calculate the percent difference.
2. Measure the voltages and currents specified in Table 6-6. Use the polarity and current direction shown on Figure 6-2. Demonstrate the I_{R1} and V_{R2} to your instructor.

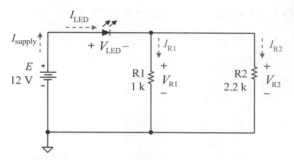

FIGURE 6-2 Series-Parallel Circuit (*pre-lab*)

Be sure you have completed Figures 6-3 and 6-4 before you proceed.

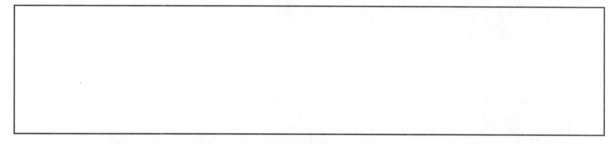

FIGURE 6-3 Schematic of Figure 6-2 Using Bubble Notation for the Source (*pre-lab*)

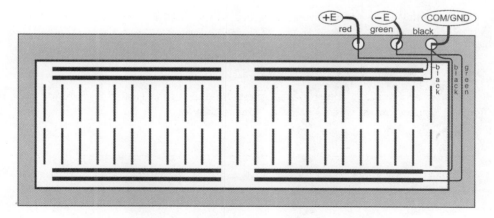

FIGURE 6-4 Intended Wiring Diagram of Figure 6-2 (*pre-lab*)

TABLE 6-5 Resistor Table for Figure 6-2

Resistor	Nominal Value	Measured Value	Percent Difference	Meets Spec?
R_1	1 kΩ			☐ Yes ☐ No
R_2	2.2 kΩ			☐ Yes ☐ No

TABLE 6-6 Data Table for Circuit of Figure 6-2

Parameter	Expected Value	Measured Value	Percent Difference
E			
V_{LED}	2 V		
V_{R1}			
V_{R2}		STOP	
I_{R1}		STOP	
I_{R2}			
I_{supply}			

STOP Instructor sign-off of I_{R1} and V_{R2} _____

SAMPLE CALCULATIONS

Show your pre-lab expected calculations (assume the LED drops 2 V) for
V_{R1} using a KVL loop or voltage walkabout on the left window (E, V_{LED}, and V_{R1})

$V_{R1} =$

V_{R2} using a KVL loop or voltage walkabout on the right window (V_{R1} and V_{R2})

$V_{R2} =$

I_{R1} using Ohm's Law: $I_{R1} =$

I_{R2} using Ohm's Law: $I_{R2} =$

I_{supply} using KCL: $I_{supply} = I_{LED} =$

TABLE 6-8 Data Table for Figure 6-5

Parameter	Expected Value	Measured Value	Percent Difference
$+E_{supply}$	+15 V		
$-E_{supply}$	−15 V		
V_{in}	5 V		
V_a			
V_{Ri}		STOP	
I_{Ri}		◆	
I_{Rf}		◆	
V_{Rf}		STOP	
V_{out}		STOP	

◆ Calculate measured currents using measured V_R and R values and Ohm's Law. Do not use an ammeter.

STOP Instructor sign-off of values in Table 6-8 _____

SAMPLE CALCULATIONS

Show your pre-lab expected calculations for:

$V_a =$

$V_{Ri} =$

$I_{Ri} =$

$I_{Rf} =$

$V_{Rf} =$

$V_{out} =$

Observations ■

1. How do the measured values compare to the expected values? Quantify values in terms of percentage differences.

   _____

   _____

2. Is the output voltage noninverted (same polarity as V_{in}) or inverted from V_{in}?   □ Noninverted □ Inverted
3. Is the output voltage greater than the input voltage?   □ Yes   □ No
4. What is the voltage gain (V_{out} / V_{in})?

   _____

Synthesis

Inverting Voltage Amplifier

The circuit of Figure 6-8 is identical to Figure 6-5 of the previous procedure except that V_{in} is applied to R_i (previously tied to common) and pin 12 is now connected to common (previously tied to V_{in}). **Make these changes with the power removed.**

Measure and record each of the parameters in Table 6-9. Demonstrate values to your instructor and calculate percent differences.

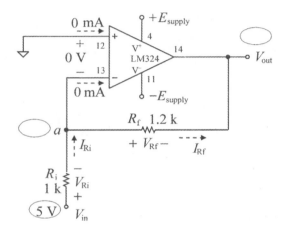

FIGURE 6-8   Inverting Voltage Amplifier

Refer to Figures 6-9 and 6-10.

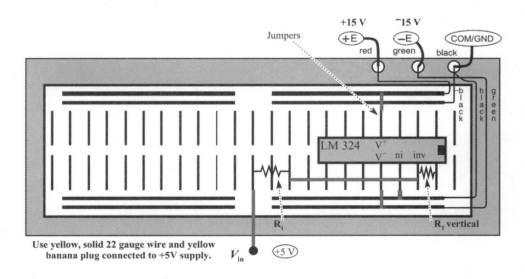

FIGURE 6-9   Wiring Board Layout for the Inverting Voltage Amplifier

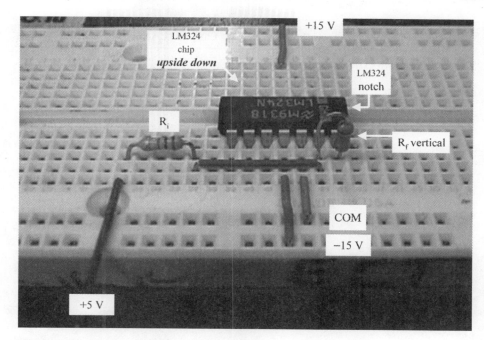

FIGURE 6-10 Photograph of the Inverting Voltage Amplifier of Figure 6-8

TABLE 6-9 Data Table for Figure 6-10

Parameter	Expected Value	Measured Value	Percent Difference
$+E_{supply}$	+15 V		
$-E_{supply}$	−15 V		
V_{in}	5 V		
V_a			
V_{Ri}		⬛STOP	
I_{Ri}		◆	
I_{Rf}		◆	
V_{Rf}		⬛STOP	
V_{out}		⬛STOP	

◆ Calculate measured currents using measured V_R and R values and Ohm's Law. Do not use an ammeter.

⬛STOP Instructor sign-off of values in Table 6-9 _____

SAMPLE CALCULATIONS

Show your pre-lab expected calculations for:

$V_a =$

$V_{Ri} =$

$I_{Ri} =$

$I_{Rf} =$

$V_{Rf} =$

$V_{out} =$

Observations

1. How do the measured values compare to the expected values? Quantify values in terms of percentage differences.

   _____

   _____

2. Is the output voltage noninverted (same polarity as V_{in}) or inverted from V_{in}?

   ☐ Noninverted   ☐ Inverted

3. Is the absolute value of the output voltage greater than the input voltage?

   ☐ Yes   ☐ No

4. What is the voltage gain (V_{out} / V_{in})?

   _____

Series Circuits

Name: _____   Date: _____

Lab Section: _____ _____   Lab Instructor: _____
                    day   time

Text Reference ▪

DC/AC Circuits and Electronics: Principles and Applications
Chapter 8: Series Circuits

Materials Required ▪

Triple power supply (2 @ 0–20 volts DC; 1 @ 5 volts DC)
1   each 100-Ω, 220-Ω, 330-Ω, 470-Ω, 560-Ω, 1-kΩ, 1.8-kΩ resistor
1   Resistor determined by student
1   each LED, red and green
1   10-kΩ single-turn potentiometer
1   2N3904 transistor

Introduction ▪

A series circuit is one with a single current path. Therefore, the current through each component is the same.
In this exercise, you will:

▪ Examine a series circuit first with a single positive supply and then with a split (positive and negative)
  supply

▪ Investigate a series circuit with two LEDs in series with a current-limiting resistor and another utilizing a
  potentiometer to control the brightness of an LED

▪ Investigate a BJT circuit.

In a BJT the collector and base currents, I_C and I_B respectively, add to form the emitter current (I_E). Since I_B is
much smaller than the I_C, the collector and emitter legs essentially form a series circuit and I_E is approximately
equal to I_C.

Pre-Lab Activity Checklist

☐ Draw the intended wiring diagrams for:
    ☐ Figure 7-1 in Figure 7-2
    ☐ Figure 7-6 in Figure 7-7
☐ Find the expected values in:

☐ Table 7-2	☐ Table 7-7
☐ Table 7-3	☐ Table 7-8
☐ Table 7-4	☐ Table 7-9
☐ Table 7-5	☐ Table 7-10

☐ Build the circuit of Figures 7-10 and 7-11 on the right end of your protoboard.
☐ Complete the sample calculations to confirm the values in Table 7-12.

Performance Checklist

☐ Pre-lab completed?   🛑 Instructor sign-off _____

☐ Table 7-4: Demonstrate measured E, V_{SW}, V_b, V_d, V_{bd}.
☐ Table 7-9: Demonstrate measured V_R, V_{POT}, V_{LED}.
☐ Table 7-12: Demonstrate measured $+E_{supply}$, $-E_{supply}$, V_B, V_E, V_a, V_C, V_{CE}.

Procedure 7-1

Total Resistance

Be sure you have completed Figure 7-2 before proceeding.

1. Measure the resistors used in Figure 7-1 and record the values in Table 7-1. Calculate the percent difference.
2. With the power supply leads removed, measure the total circuit resistance with the switch closed and open and record those values in Table 7-2. *This is the proper way to measure total resistance.*
3. Set two supply voltages to 6.5 V, then connect them to the circuit.
4. Turn off the supply and measure the total circuit resistance with the switch open and closed and record those values in Table 7-2. *This is **NOT** the proper way to measure total resistance.*

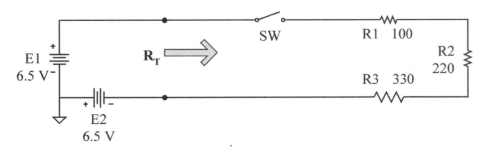

FIGURE 7-1   Series Circuit: Resistance Measurements

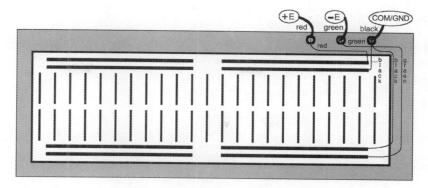

FIGURE 7-2   Intended Wiring Diagram of Series Circuit of Figure 7-1 (*pre-lab*)

TABLE 7-1   Resistor Table for Figure 7-1

Resistor	Nominal Value	Tolerance %	Measured Value	Percent	Meet Specs?
R_1	100 Ω				☐ Yes ☐ No
R_2	220 Ω				☐ Yes ☐ No
R_3	330 Ω				☐ Yes ☐ No

TABLE 7-2   Total Series Circuit Resistance for Figure 7-1

Parameter	Expected Value	Measured Value	Percent Difference
R_T with SW **closed** (supply removed)			
R_T with SW **open** (supply removed)			
R_T with SW **closed** (supply connected)			
R_T with SW **open** (supply connected)			

Note: R_T is the total resistance as viewed from the supply, including SW, R_1, R_2, and R_3.

SAMPLE CALCULATIONS

Show your sample pre-lab calculations for:

R_T (switch closed) =

R_T (switch open) =

Observations

1. Using the resistor tolerances, find the worst case maximum and minimum total resistance values with the switch closed. Show your work.

   $R_{T(max)} =$ _____

   $R_{T(min)} =$ _____

2. What is your measured total resistance R_T percent difference with the switch closed? Show your work.

   % Diff = _____

3. Which ohmmeter range should be used to measure an **open**?   ☐ Lowest ☐ Highest

4. Compare the total resistance readings with and without the supply connected. What do you observe?

   _____

5. When measuring with the supply connected, does the reading increase or decrease from the correct reading?   ☐ Increase ☐ Decrease

Procedure 7-2

Series Circuit Current

1. Turn on the supply voltages previously set to 6.5 V each.

2. Measure the supply current and the current going *into* each resistor and verify that the current is the same throughout a series circuit. See Figure 7-3. Record the measurements in Table 7-3.

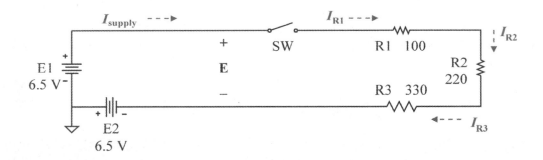

FIGURE 7-3 Series Circuit: Current Measurements *with Switch Closed* (same as Figure 7-1)

TABLE 7-3 Series Circuit Current with Switch Closed (Data Table for Figure 7-3)

Parameter	Expected Value	Measured Value	Percent Difference
E_1	6.5 V		
E_2	6.5 V		
E			
I_{supply}			
I_{R1}			
I_{R2}			
I_{R3}			

SAMPLE CALCULATIONS

Show your sample pre-lab calculations for:

$I_{supply} =$

$I_{R1} =$

Observations ▪

1. From your measurements, is the series current the same at each place? ☐ Yes ☐ No
2. How close were the measured values to the expected values? Quantify based upon percentage errors.

_____

_____

Procedure 7-3

Series Circuit Voltages with a Closed Switch

1. Using the circuit of Figure 7-4, measure the voltage drops and node voltages. Record the measurements in Table 7-4.

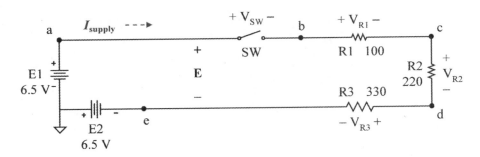

FIGURE 7-4 Series Circuit: Voltage Measurements *with Switch Closed* (same as Figure 7-1)

TABLE 7-4 Series Circuit Voltages *with Switch Closed* (Data Table for Figure 7-4)

Parameter	Expected Value	Measured Value	Percent Difference
E		⬤STOP	
V_{SW}		⬤STOP	
V_{R1}			
V_{R2}			
V_{R3}			
V_a			
V_b		⬤STOP	
V_c			
V_d		⬤STOP	
V_e			
V_{bd}		⬤STOP	

⬤STOP Instructor sign-off of measured values in Table 7-4 _____

SAMPLE CALCULATIONS

Using the voltage divider rule, show your sample pre-lab calculations for:

$V_{R1} =$

$V_{bd} =$

Observations

1. What is the voltage drop across the closed switch?_____

2. Use Ohm's Law with measured values (V_{SW} and I_{supply}) to calculate the resistance of the switch.

   $R_{SW} =$ _____

Procedure 7-4

Series Circuit Voltage with Switch Open

1. Using the circuit of Figure 7-5 (switch open), measure the voltage drops and node voltages. Record the measurements in Table 7-5.

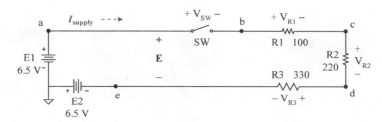

FIGURE 7-5 Series Circuit: Voltage Measurements with Switch Open (same as Figure 7-1)

TABLE 7-5 Series Circuit Voltages *with Switch Open* (Data Table for Figure 7-4)			
Parameter	**Expected Value**	**Measured Value**	**Percent Difference**
E			
I_{supply}			
V_{SW}			
V_{R1}			
V_{R2}			
V_{R3}			
V_a			
V_b			
V_c			
V_d			
V_e			

SAMPLE CALCULATIONS

Using KVL, show your sample pre-lab calculations for:

$V_{SW} =$

Observations ◼

1. What is the current in an open circuit? _____

2. What is the voltage drop across your open switch? _____

3. A **single open** in a series circuit drops how much voltage? _____

4. Based upon the **node voltage measurements,** suggest a technique to locate an open in a series circuit.

_____

_____

Procedure 7-5
Current-Limiting Circuit to Protect Two LEDs

PRE-LAB CALCULATIONS

Determine the voltage drop across the resistor of Figure 7-6.

$V_R =$

Using the calculated V_R, find the resistor value, R, needed to limit the LED current to a maximum of 10 mA. Select a standard resistance value that produces a current as close as possible to 10 mA *without exceeding 10 mA*. Show your work.

Using your selected resistance value, calculate:

$I_R =$

Be sure to complete Figure 7-7 before proceeding.

1. Measure your selected value for R and record the value in Table 7-6. Note that improper selection of the resistor, R, may damage to the LEDs.

TABLE 7-6 Resistor Data Table for Figure 7-6

Resistor	Nominal Value	Measured Value	Percent Difference	Meet Specs?
R				☐ Yes ☐ No

2. Build and apply power to the circuit of Figure 7-6. Are both LEDs lit? ☐ Yes ☐ No

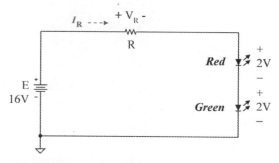

FIGURE 7-6 Two-LED Circuit

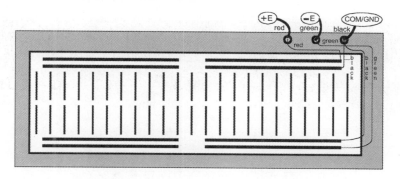

FIGURE 7-7 Intended Wiring Diagram for Circuit of Figure 7-6 (*pre-lab*)

3. Measure and record the values specified in Table 7-7. Improper measurement techniques may damage the LEDs.

TABLE 7-7 Data Table for Figure 7-6

Parameter	Expected Value	Measured Value	Percent Difference
E	16 V		
Red V_{LED}			
Green V_{LED}			
V_R			
I_R			

4. Reverse the red LED. Are the LEDs lit? ☐ Yes ☐ No

   ⊐ **Note:** Some LEDs may have a reverse breakdown voltage of 10 to 12 volts. This may result in enough current to turn on the green LED.

5. Measure and record the values specified in Table 7-8 on page 90.

TABLE 7-8 Data Table for Figure 7-6 with the RED LED Reversed

Parameter	Expected Value	Measured Value	Percent Difference
E	16 V		
Red V_{LED}		◆	
Green V_{LED}			
V_R			
I_R			

◆ The DMM DC voltmeter has a resistance of about 10 MΩ and completes the circuit. This reading may be less than the expected value.

Observations

1. Did your designed value of resistance, R, provide a current close to but not exceeding 10 mA?

   ☐ Yes   ☐ No

2. Which component(s) are most likely to have the highest percent difference values? (See Table 7-7).

   _____

3. How do the component variations affect the design?

   _____

   _____

4. What was the effect of reversing the red LED on circuit current?

   _____

5. Does the measured voltage across the reversed red LED make sense?   ☐ Yes   ☐ No

   Why or why not? _____

   _____

   _____

Procedure 7-6
Light Dimmer Switch

Be sure to complete Figure 7-9 before proceeding.

1. With the single-turn potentiometer fully clockwise, measure the resistance between the outer terminals, the left and middle terminals, and the right and middle terminals.

2. Build the circuit of Figure 7-8 and measure R_{POT} and $R + R_{POT}$. Adjust $R + R_{POT}$ to its maximum value and record the values in Table 7-9. Demonstrate those values to your instructor.

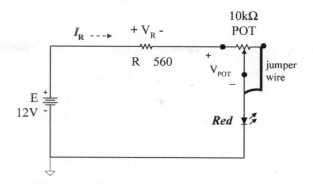

FIGURE 7-8 LED Light Dimmer Circuit

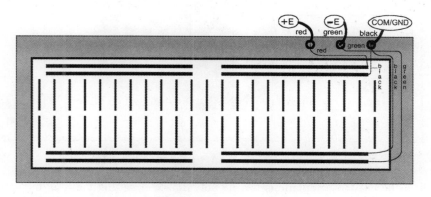

FIGURE 7-9 Intended Wiring Diagram of Figure 7-8 (use the 1-turn potentiometer)

TABLE 7-9 Minimum LED Current

Parameter	Expected Value	Measured Value	
R_{POT}	10 kΩ		
$R + R_{POT}$			
E	12 V		
I_{min}			STOP
LED light	**DIM**		
V_R			STOP
V_{POT}			STOP
V_{LED}			STOP

STOP Instructor sign-off of measured values in Table 7-9 _____

3. Adjust R + R_{POT} to its minimum value and record the values in Table 7-10.

TABLE 7-10 Maximum LED Current

Parameter	Expected Value	Measured Value
R_{POT}	0 Ω	
$R + R_{POT}$		
E	12 V	
I_{max}		
LED light	**BRIGHT**	
V_R		
V_{POT}		
V_{LED}		

SAMPLE CALCULATIONS

Show the pre-lab calculations for the potentiometer setting yielding the expected minimum current (R_{POT} = 10 kΩ).

$R + R_{POT}$ =

I_{MIN} =

V_R =

V_{POT} =

Show the pre-lab calculations for the potentiometer setting yielding the expected maximum current (R_{POT} = 0Ω.

$R + R_{POT}$ =

I_{MAX} =

V_R =

V_{POT} =

Observations

1. Calculate the dim LED *static resistance.*

$$R_{LED} = \frac{V_{LED}}{I_{LED}} = \frac{V_{dim}}{I_{dim}} = \underline{\hspace{6cm}}$$

2. Calculate the bright LED *static resistancem.*

$$R_{LED} = \frac{V_{LED}}{I_{LED}} = \frac{V_{bright}}{I_{bright}} = \underline{\hspace{6cm}}$$

3. Calculate the LED *dynamic resistance.* $\quad \dfrac{\text{change in } V}{\text{change in } I}$

$$R_{LED} = \frac{\Delta V_{LED}}{\Delta I_{LED}} = \frac{V_{bright} - V_{dim}}{I_{bright} - I_{dim}} = \underline{\hspace{6cm}}$$

Synthesis

BJT Circuit Voltages and Currents

Measure the resistors for Figure 7-10 and record the values in Table 7-11. Build the circuit of Figure 7-10, then measure and record the values specified in Table 7-11.

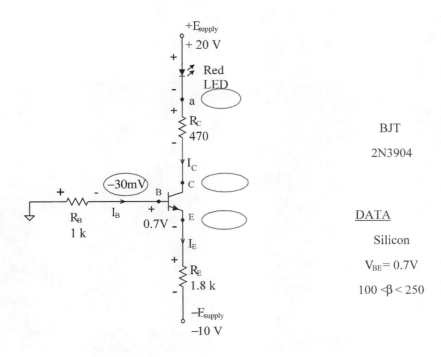

FIGURE 7-10 BJT Circuit (*pre-lab*)

TABLE 7-11 Resistor Data Table for Figure 7-10

Resistor	Nominal Value	Measured Value	Percent Difference	Meet Specs?
R_C	470 Ω			☐ Yes ☐ No
R_B	1 kΩ			☐ Yes ☐ No
R_E	1.8 kΩ			☐ Yes ☐ No

Refer to the the BJT circuits shown in Figures 7-11 and 7-12 and complete Table 7-12.

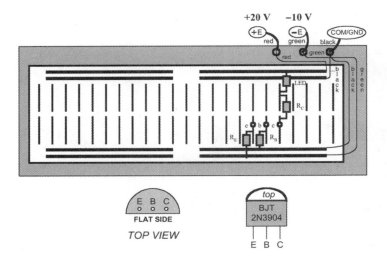

FIGURE 7-11 BJT Circuit Board Layout (*pre-lab*)

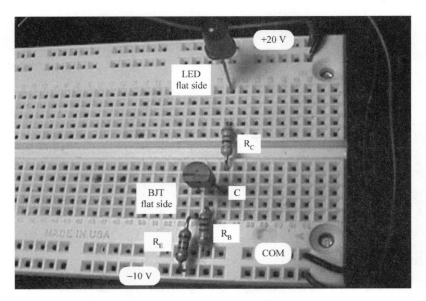

FIGURE 7-12 Photograph of Circuit Layout (*pre-lab*)

TABLE 7-12 Data Table for Figure 7-10

Parameter	Expected Value	Measured Value	Percent Difference
$+E_{supply}$	+20 V	STOP	
$-E_{supply}$	−10 V	STOP	
$^*V_B$	≅ − 30 mV	STOP	
$^*V_E$	− 0.73 V	STOP	
$^*V_a$	18 V	STOP	
$^*V_C$	15.58 V	STOP	
V_{CE}	16.31 V	STOP	
V_{BE}	0.7 V		
V_{LED}	2 V		
V_{RB}	≅ − 30 mV		
V_{RE}	9.27 V		
V_{RC}	2.42 V		

*Fill in node voltages (in the ovals) on the schematic of Figure 7-10.

STOP Instructor sign-off of measured values in Table 7-12 _____

SAMPLE CALCULATIONS

Using V_{BE} and V_B provided on Figure 7-10, complete the calculations below and confirm the expected values in Table 7-12.

$V_E = V_B - V_{BE} =$

$V_{RE} = V_E - (-E_{supply} =$

$I_C \cong I_E = \dfrac{V_{RE}}{R_E} =$

$V_{RC} = I_C \times R_C =$

$V_a = +E_{supply} - V_{LED} =$

$V_C = V_a - V_{RC} =$

$$V_{CE} = V_C - V_E =$$

MINIMUM $\quad I_B = \dfrac{I_E}{\beta_{max}} =$

MINIMUM $\quad V_{RB} = I_B \times R_B =$

MAXIMUM $\quad I_B = \dfrac{I_E}{\beta_{min}}$

MAXIMUM $\quad V_{RB} = I_B \times R_B =$

Observations ■

1. How did your measured values compare with the expected values?

_____

2. Calculate your measured currents using Ohm's Law for resistance.

$$I_B = I_{RB} = \dfrac{V_{RB}}{R_B} = \underline{\hspace{10cm}}$$

$$I_E = I_{RE} = \dfrac{V_{RE}}{R_E} = \underline{\hspace{10cm}}$$

$$I_C = I_{RC} = \dfrac{V_{RC}}{R_C} = \underline{\hspace{10cm}}$$

3. Write the KCL expression for the BJT currents just found.

_____

4. Do these BJT currents satisfy Kirchhoff's Current Law? □ Yes □ No

5. Although I_C and I_E may not be exactly the same, are they close enough to assume that R_C and R_E are essentially in series for this particular BJT circuit and its circuit values?

□ Yes □ No

6. Using the I_C and I_B found above, calculate the measured β of your BJT.

$$\beta = \dfrac{I_C}{I_B} = \underline{\hspace{10cm}}$$

Oscilloscope, Thévenin Analysis, and Voltage Division

Name: _____ Date: _____

Lab Section: _____ _____ Lab Instructor: _____
day   time

Text Reference ◼

DC/AC Circuits and Electronics: Principles and Applications
Chapter 8: Series Circuits

Materials Required ◼

Triple power supply (2 @ 0–20 volts DC; 1 @ 5 volts DC)
Oscilloscope
1   each 330-Ω, 470-Ω, 820-Ω, 1-kΩ, 1.2-kΩ, 1.8-kΩ, 2.2-kΩ and 3.3-kΩ resistor
1   10-kΩ multi-turn potentiometer
1   2N3904 transistor

Introduction ◼

In this exercise, you will:

- Observe positive and negative DC voltages using the oscilloscope.
- Learn how to choose a proper V/DIV setting to achieve an optimal oscilloscope display.
- Experimentally derive a Thévenin circuit model.

Thévenin's theorem states that *any linear bilateral network may be modeled as an ideal voltage source in series with a resistance* (E_{TH} and R_{TH}, respectively). By measuring the open circuit voltage and short circuit current of a resistive network, one can experimentally determine the equivalent Thévenin circuit model. The open circuit voltage ($V_{OC} = E_{TH}$) is found by replacing the circuit load with a voltmeter. The short circuit current (I_{SC}) is found by replacing the circuit load with an ammeter. The Thévenin resistance is then found using the formula:

$$R_{TH} = V_{OC}/I_{SC} = E_{TH}/I_{SC}$$

In the last exercise you learned that the base current (I_B) in a BJT is much smaller than the current in the collector (I_C) or emitter (I_E). In order to set the voltage on the base (known as *biasing* the base) a *voltage divider* is used. It consists of two resistors connected to a voltage with their common point connected to the base. If I_B is much smaller than the currents through the two resistors, it can be ignored, making the two-resistor biasing circuit essentially a series circuit. This allows us to utilize the voltage divider rule (VDR) to bias the base voltage.

Pre-Lab

Pre-Lab Activity Checklist

☐ Perform pre-lab calculations to find E_{TH} in Procedure 8-5.

☐ Find the expected values in Table 8-7 including sample calculations.

☐ Build the circuits of Figures 8-5 and 8-7.

Performance Checklist

☐ Pre-lab completed?   🛑 Instructor sign-off _____

☐ Procedure 8-2: Demonstrate 3.4-V measurement.

☐ Table 8-2: Demonstrate measured − 4.6 V.

☐ Table 8-5: Demonstrate measured E_{TH}, I_{SC}, V_{POT}, and R_{POT}.

☐ Table 8-7: Demonstrate measured V_c and V_{ce} with both the DMM and oscilloscope.

Procedure 8-1

Introduction to the Oscilloscope

The oscilloscope (or "scope") is essentially a visual waveform voltmeter that displays DC voltages, AC voltages (alternating voltages), and combined AC and DC voltages as a function of time. The image (or "trace") of a signal is projected onto a display such as a cathode-ray tube (CRT) or liquid crystal display (LCD) equivalent to the display in a computer monitor or television.

The Tektronix 2213 display used in the example photo shown in Figure 8-1 has an 8 cm × 10 cm graticule display with which to take visual measurements of voltages. These 1 cm divisions are further subdivided into one fifth of a cm. The precision of the reading is limited, by the accuracy of the observer's view and by the size of the display, to typically two or three significant figures.

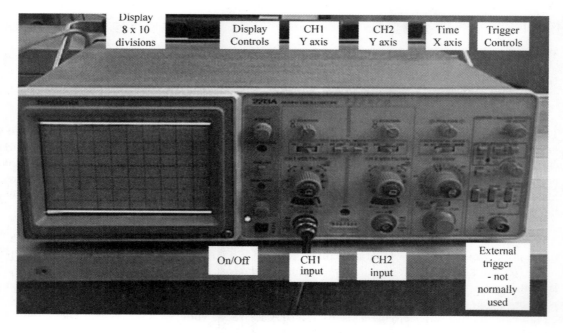

FIGURE 8-1 Photograph of Tektronix 2213 Oscilloscope

Most modern oscilloscopes have two or more "channels" or inputs. Channels control the vertical (y) deflection of the trace. These channels may be labeled 1, 2, or A, B, etc. They have controls in volts/division (V/DIV) with various ranges. Each channel has a position control that allows the user to choose the location of a reference (GND). Each channel also has a selector to choose DC, AC, or GND (ground) coupling. With DC coupling, both AC and DC portions of a waveform are displayed. With AC coupling, any DC portion of a signal is removed and only the AC is displayed. The GND coupling position applies 0 V to the channel regardless of the signal connected to the input. Using the position control with GND coupling allows the user to choose the vertical position of the 0-V reference on the display.

Channels are connected to circuits through "probes," which typically come in 1x and 10x varieties. A 1x probe applies the signal directly to the input so that if 100 V are applied to the probe, the input will see the entire 100 V. A 10x probe reduces the signal applied to the input by one-tenth so that, if 100 V are applied to the probe, the input sees only 10 V. When using a 10x probe you may need to multiply the V/DIV scale by ten (thus the name 10x) in order to compensate for the probe. Modern oscilloscopes can detect a 10x probe and will display the proper scale, but older models require the user to manually multiply the scale value by ten.

The horizontal (x) deflection of the trace is controlled by a sweep generator. The sweep generator moves the beam repeatedly from left to right. It has a sweep-rate control labeled in seconds/division (SEC/DIV). If the sweep rate is set to 0.5 SEC/DIV, the trace moves one division in one-half second. Since there are ten horizontal divisions on the display, it will take 5 seconds for the beam to cross the screen from left to right. As the sweep rate increases, the beam moves more quickly and eventually appears as a solid line.

The initiation of the left-to-right movement of the beam is controlled by the trigger circuitry. Depending upon the trigger settings, the beam will not commence until the input voltage reaches a certain "triggering level." Since we are working with only DC levels in this exercise, triggering should not be a problem; but if the trace disappears, you may have a triggering misadjustment.

Since oscilloscope designs vary widely, you may need a brief demonstration from your instructor on how to use the controls for your particular scope.

Understanding oscilloscope operation is critical. Ask questions!

1. Connect a 1x oscilloscope probe to Channel 1 (or A) of your scope. Do not connect anything to the probe. Turn the oscilloscope on.

2. Set the SEC/DIV control to 200 mS and adjust the SEC/DIV calibration knob to its calibrated position.

3. Set the V/DIV of Channel 1 to 0.5 V and adjust the V/DIV calibration knob to its calibrated position. Set the channel selection control to display Channel 1.

4. Adjust the ↕ position knob for Channel 1 to center the trace on the display. You should see the beam crossing the screen about once every two seconds.

5. Using the focus control, adjust the beam so that is as sharp as possible. Adjust the brightness control to get a beam that is plainly visible.

6. Set the SEC/DIV knob to 1 mS, then adjust the ↔ position knob to center the trace on the display. You should now see a solid line that crosses the entire grid of the display.

7. Set the input coupling of Channel 1 to the ground (GND) position. Selecting GND coupling applies 0 V to the input of the scope regardless of what is connected to the probe.

8. Adjust the ↕ position knob to move the trace to the center of the display. Positioning the trace while using GND coupling allows you to choose the 0-V reference position.

9. Set the V/DIV of Channel 1 to 1 V and adjust the V/DIV calibration knob to its calibrated position.

10. Connect the probe to the power supply as shown in Figure 8-2 on page 100. If you are using a clip-type probe, connect the red lead to the positive terminal on the supply.

11. Set the input selector of Channel 1 to the DC position.

12. Set the power supply to the minimum position, then turn it on.

13. Slowly increase the supply voltage while observing the display. It should move up the display. If you increase the voltage too much it will go off screen.

14. Set the voltage so that the trace follows the top line of the display grid, then measure the supply voltage with the DMM and record that value here:

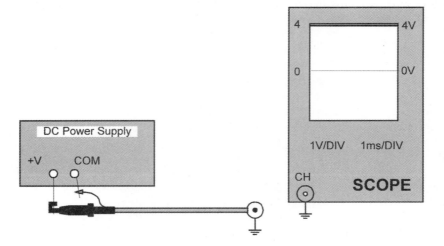

FIGURE 8-2 Observing a DC Signal on the Scope

> ◲ **Note:** The trace is 4 divisions (cm) above the reference (0 V) at the center. Since Channel I is set at I V/DIV the indicated voltage should be approximately 4 V.

15. Without changing the supply voltage (~4 V), reverse the probe leads on the power supply and note the position of the trace.

_____

16. Take the Channel 1 calibration knob out of calibration and observe the result.

_____

Observation ■

1. If the channel is out of calibration, can you use the scope to accurately measure voltages?

☐ Yes   ☐ No

Procedure 8-2

DC Voltage Measurements with a Scope

1. Leave the DC supply set to 4 V, but return the supply leads to the original position shown in Figure 8-2.

2. Set the input selector to GND and adjust the ↕ position knob to move the trace to the very bottom of the display grid.

> ◲ **Note:** Since we are displaying a positive voltage, we can sometimes improve the reading by allowing more display "room."

3. Return the V/DIV calibration knob to its calibrated position and make sure that the V/DIV of Channel 1 is at 1 V.

4. Set the input-coupling selector of Channel 1 to DC. Does the scope still display 4 V?

        ☐ Yes ☐ No

5. Change the V/DIV of Channel 1 to 0.5 V. Does the scope still display 4 V?

        ☐ Yes ☐ No

6. Turn off the DMM and set the voltage to 3.4 V using the oscilloscope. Which scale is most precise?

        ☐ 1 V/DIV ☐ 0.5 V/DIV

7. Measure the supply voltage with the DMM and record that value here: _____

🛑 Demonstrate the 3.4-V scope and DMM reading to your instructor _____

8. Take Channel 1 out of calibration. Does the DMM reading change?

        ☐ Yes ☐ No

9. Return Channel 1 to calibration before proceeding.

Observations ■

1. Which is more precise, the scope or the DMM?   ☐ Scope ☐ DMM

   Why? _____

2. How many significant digits can one read off the display?

   ☐ 1 ☐ 2 ☐ 3 ☐ 4 ☐ > 4

3. If you wanted to read a negative voltage, where would you place the 0-V (GND) reference?

   ☐ Top ☐ Middle ☐ Bottom

4. If the channel is out of calibration, can you use the scope to accurately measure voltages?

        ☐ Yes ☐ No

Procedure 8-3
Choosing V/DIV on the Scope

We will now take you through the formal method for choosing a proper scale for the oscilloscope display. The first voltage displayed will be 15 V. Afterward you will display both positive and negative DC voltages.

◢ **Note:** In later exercises, you will use the oscilloscope to view AC waveforms, so proper selection of input sensitivity (V/DIV) is important.

SAMPLE CALCULATIONS

Choice of V/DIV scale

Since the display is 8 divisions high, we want to display our 15 volts over the as much of the display as possible. If we could spread 15 V over the entire 8 vertical divisions we would get

15 V/8 divisions = 1.875 V/DIV

There is no sensitivity of 1.875 V/DIV. Sensitivity selections are typically multiples of 1, 2, or 5 (0.1 V, 0.2 V, 0.5 V, 1 V, 2 V, 5 V, etc.). In order to display 15 V, we must go to the next *higher* sensitivity.

1.875 V/DIV2 → V/DIV

To determine how many divisions 15 V will be offset from the reference,

15 V/(2V/DIV) = 7.5 divisions

> ⊐ **Note:** Had we used the next lower scale (1 V/DIV), it would have required 15 divisions to display.

---

1. Using the information determined above, use the oscilloscope to set the power supply to 15 V. **Do not use the DMM**.

2. Complete a sketch of the display in Figure 8-3 and fill in all specified values including vertical and horizontal axes settings and 0-V (GND) position. Indicate actual vertical axis voltage values relative to GND. Indicate actual horizontal axis time values starting with 0 mS.

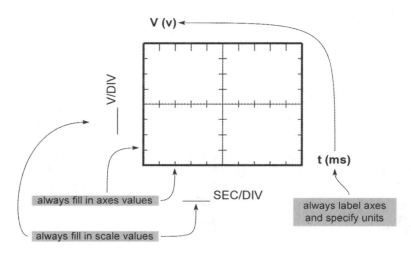

FIGURE 8-3 _____
Title (Always give your figures a number and a title)

3. Measure the supply voltage with the DMM and complete Table 8-1.

TABLE 8-I DC Voltage Measurement with DMM and Scope

Parameter	Expected Value	Measured Value	Percent Difference
E using Scope*	15 V		
E using DMM	15 V		

*1x indicator with 1x probe or cable

Procedure 8-4

Setup of a Negative Voltage Using the Scope

1. Calculate the proper sensitivity to display − 4.6 V.

SAMPLE CALCULATIONS

Show the calculations for selection of the V/DIV setting to display −4.6 V. Follow the example in Procedure 8-3.

---

2. Complete a sketch of the display in Figure 8-4, filling in all specified values .

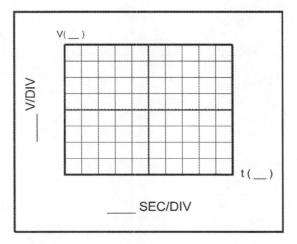

FIGURE 8-4 _____

3. Measure the supply voltage with the DMM and complete Table 8-2 on page 104.
4. Turn off the oscilloscope

TABLE 8-2 DC Voltage Measurement with DMM and Scope

Parameter	Expected Value	Measured Value	Percent Difference
E using Scope*	− 4.6 V	(STOP)	
E using DMM	− 4.6 V	(STOP)	

*Ix indicator with Ix probe or cable

(STOP) Demonstrate the − 4.6-V scope and DMM reading to your instructor _____

Observations

1. Were the DMM and scope readings reasonably close? ☐ Yes ☐ No
2. Can the scope be used to measure voltage directly? ☐ Yes ☐ No
3. Can the scope be used to measure current directly? ☐ Yes ☐ No

Procedure 8-5

Experimental Derivation of a Thévenin Model

Refer to Figure 8-5.

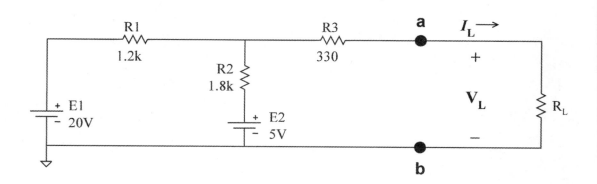

FIGURE 8-5 Circuit Under Investigation to Find Thévenin Model and Load Values (*pre-lab*)

SAMPLE CALCULATIONS

Perform circuit analysis to find E_{TH} of the circuit shown in Figure 8-5. Remove the load (R_L) and determine the voltage across the load's open terminals (V_{ab}). Show your work.

Your value should agree with the expected value in Table 8-4.

1. Measure the resistors used in Figure 8-5 including the three values of fixed resistor for R_L. Record the values in Table 8-3. Calculate the percent difference.

TABLE 8-3 Resistor Data for Figure 8-4

Resistor	Nominal Value	Measured Value	Percent Difference	Meet Specs?
R_1	1.2 kΩ			☐ Yes ☐ No
R_2	1.8 kΩ			☐ Yes ☐ No
R_3	330 Ω			☐ Yes ☐ No
R_{L1}	470 Ω			☐ Yes ☐ No
R_{L2}	1 kΩ			☐ Yes ☐ No
R_{L3}	2.2 kΩ			☐ Yes ☐ No

2. Set two supply voltages, E_1 and E_2.
3. Connect the power supply to your circuit paying close attention to common. Leave R_L out of the circuit initially.
4. Measure the Thévenin voltage (no load or open-circuit voltage) and record the value in Table 8-4 on page 106.

TABLE 8-4   Model Parameter Data for Figure 8-4

Parameter	Expected Value	Measured Value (DMM)
$E_{TH} = V_{OC} = V_{NL} = V_{ab \text{ without load}}$	14 V	
V_L with R_{L1}	4.33 V	
V_L with R_{L2}	6.83 V	
V_L with R_{L3}	9.48 V	
I_L with R_L a short (I_{SC})	13.33 mA	

5. Measure the load voltage for each of the specified resistances and record the value in Table 8-4.

6. Measure the short circuit current (I_{SC}) and record the value in Table 8-4. The ammeter is essentially a short, so you may connect the ammeter from point a to point b to measure I_{SC}.

Observations   ■

(using measured values)

1. Use the measured values of V_{OC}, V_{RL1}, and R_{L1} to determine and draw the Thévenin model of this circuit.

2. Using your Thévenin model, *calculate* the load voltage V_{RL2} with R_{L2} attached. How does this compare with your measured V_{RL2} value (percent difference)?

   $V_{RL2} = $ _____ ; _____

3. Using your Thévenin model, *calculate* the load voltage V_{RL3} with R_{L3} attached. How does this compare with your measured V_{RL3} value (percent difference)?

   $V_{RL3} = $ _____ ; _____

4. Using your Thévenin model, *calculate* I_{SC}, the short circuit current (load is a short). How does this compare with your measured I_{SC} value (percent difference)?

   $I_{SC} = $ _____ ; _____

5. Does the Thévenin model accurately predict load current and voltage?   □ Yes □ No

Procedure 8-6

Experimental Determination of R_{TH}

1. Re-measure the Thévenin voltage to assure that it has not changed.
2. Connect the multi-turn 10-kΩ potentiometer as R_L, as shown in Figure 8-5.
3. Adjust the potentiometer until V_L is one-half of E_{TH}.
   Note: This is called the "matched load" technique for finding the Thévenin resistance of a circuit.
4. Remove the potentiometer and record the resistance value of R_{POT} in Table 8-5. Is the value approximately the same as the value found in Procedure 8-5? ☐ Yes ☐ No

TABLE 8-5 Model Parameter Data (Data Table for Figure 8-5)

Resistor	Measured Value
R_{POT}	STOP

Refer to Figure 8-6.

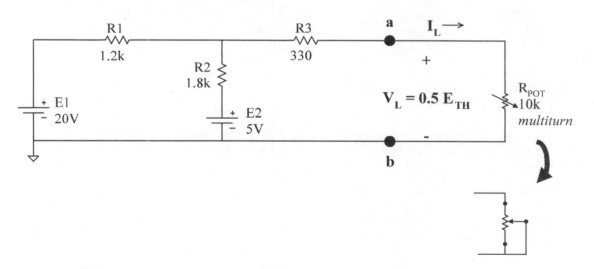

FIGURE 8-6 Circuit Under Investigation to Find Thevenin Resistance (same as Figure 8-5)

(STOP) Instructor sign-off of measured values E_{TH}, I_{SC}, V_{POT}, and R_{POT} _____

Observation ■

1. Using the Thévenin model, prove below that if the load resistance is equal to the Thévenin resistance, then the load voltage will be equal to ½ E_{TH}.

Synthesis

Transistor Biasing Circuit Using a Voltage Divider

In this exercise, the resistors R_1 and R_2 of Figure 8-7 essentially form a series-circuit voltage divider. Since the base current (I_b) of the transistor is much smaller than the currents through R_1 and R_2, we may neglect I_b to get an approximation of the base voltage (V_b). After your initial calculations you will check to assure that your assumption that I_b is much smaller than I_{R1} is valid.

Once V_b is found, we will approximate V_{be} at 0.7 V and walk down to the emitter (e) to calculate V_e. Knowing V_e we can then find I_e, which we will assume to be approximately equal to I_c (since $I_b < I_c$). We can then use Ohm's Law to find V_{RC} and then V_e and V_{ce}.

SAMPLE CALCULATIONS

Show your pre-lab calculations for:

V_{R2} (use VDR) =

$V_b =$

$V_{R1} =$

$I_{R1} \approx I_{R2} =$

$V_{be} = 0.7$ V

$V_e =$

$V_{RE} =$

$I_E =$

$I_c =$

$V_{RC} =$

$V_c =$

$V_{ce} =$

MINIMUM $I_b = I_c/\beta_{max} = I_c/250 =$

MAXIMUM $I_b = I_c/\beta_{min} = I_c/100 =$

Is the circuit β independent? (Is $I_b < I_{R1}$?)   ☐ Yes ☐ No

For the circuit of Figure 8-7, measure the resistors used and complete Table 8-6. Connect the 20-V supply and measure all values specified in Table 8-7 using the DMM. The values should be close to the expected values. Measure all values specified in Table 8-7 using the oscilloscope. Make sure to follow the instructions for measuring V_{ce}. Refer to Figures 8-8 and 8-9.

TABLE 8-6   Resistor Data Table for Figure 8-6

Resistor	Nominal Value	Measured Value	Percent Difference	Meet Specs?
R_1	3.3 kΩ			☐ Yes ☐ No
R_2	1 kΩ			☐ Yes ☐ No
R_c	820 Ω			☐ Yes ☐ No
R_e	470 Ω			☐ Yes ☐ No

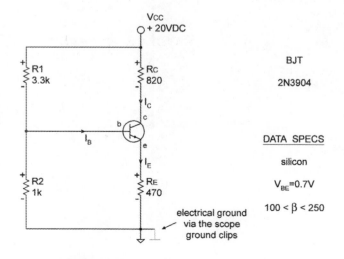

FIGURE 8-7 BJT Voltage Divider Biasing Circuit (*pre-lab*)

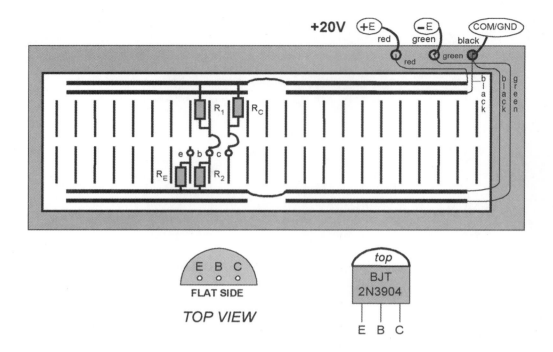

FIGURE 8-8 Suggested Wiring Board Layout: Note that the BJT flat side is facing toward the bottom of the board. Connect the BJT as indicated: **e**, **b**, and **c** (emitter, base, and collector, respectively)

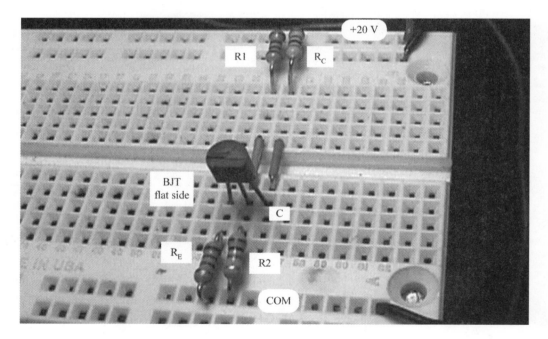

FIGURE 8-9 Photograph of Layout for BJT Circuit (*pre-lab*)

TABLE 8-7 Voltage Divider Bias (Data Table for Figure 8-7)

Parameter	Expected Value	DMM Measured Value	SCOPE Measured Value
V_{CC}			
I_{R1}			
I_{R2}			
V_b			
V_e			
I_e			
I_c			
V_{RC}			
V_c		🛑	🛑
V_{ce}		🛑	*🛑

*Do not connect the oscilloscope electrical common to the emitter. It may destroy the transistor! Follow these steps to measure V_{ce}:**

1. The ground leads on all oscilloscope probes must be connected to the circuit common. To measure between two non-common points using an oscilloscope, one must connect channel 1 to the first point (CH1 to V_c) and channel 2 to the second point (CH2 to V_e).

2. To get V_{ce} you must subtract V_e from V_c ($V_c - V_e = V_{ce}$; i.e. CH1 − CH2). To accomplish this *set both channel's V/DIV to the same scale*. Choose a scale where both channels can be observed.

3. Find the control for Channel 2 Invert and place it in the *Invert* position. This inverts Channel 2 (displays minus CH2).

4. Find the control for Alt (Alternate), Chop, or Add. Place this control in the *Add* position. This adds Channel 1 to the inverted Channel 2 (CH1 + (−CH2)).

5. Place *both* inputs in the GND coupling position and use one of the channel position controls to adjust the position of the 0-V reference.

6. Return both channels to DC coupling and observe V_{ce}.

7. You may be able to improve the resolution by setting *both* scales to a smaller V/DIV setting. (Older oscilloscopes require *both* channels to use the same V/DIV setting in order to properly add the signals.)

🛑 Instructor sign-off of V_c and V_{ce} measurements_____

Op-Amp Comparators and Series Circuits

Name: _____   Date: _____

Lab Section: _____ _____   Lab Instructor: _____
             day   time

Text Reference

DC/AC Circuits and Electronics: Principles and Applications
Chapter 9: Essentially Series

Materials Required

Triple power supply (2 @ 0–20 volts DC; 1 @ 5 volts DC)
Oscilloscope
3   1-kΩ resistors
2   2.7-Ω resistors
1   each 10-kΩ, 100-kΩ, and 470-kΩ resistors
1   unknown resistor (value determined in pre-lab)
1   10-kΩ multi-turn potentiometer
1   LM324 operational amplifier (op amp)
1   each LED, red and green

Introduction

In this exercise, you will:

- Design a voltage reference circuit using a voltage divider.

This reference voltage will be used to apply a fixed voltage to several different comparator circuits you will study. The comparator circuits will be operated in open loop and positive feedback modes. With positive feedback, an essentially series circuit is observed and the voltage applied to the input of the op amp can be determined using voltage division.

Pre-Lab Activity Checklist

☐ Perform pre-lab calculations for:

    ☐ Procedure 9-1 to determine the expected values in Tables 9-1 and 9-2

    ☐ Procedure 9-2 to determine the expected values in Tables 9-4, 9-5, and 9-6

    ☐ Procedure 9-3 to determine the expected values in Table 9-7

    ☐ Procedure 9-4 to determine the expected values in Tables 9-9, 9-10, 9-11, and 9-12

    ☐ Synthesis procedure to determine the expected values in Table 9-13.

☐ Build the circuits of Figures 9-1 and 9-2.

Performance Checklist

☐ Pre-lab completed?

☐ Demonstrate trip points in Table 9-7.

☐ Demonstrate trip points in Table 9-13.

🛑 Instructor sign-off _____

Procedure 9-1

Designing a Reference Voltage

SAMPLE CALCULATIONS (*pre-lab*)

The circuit of Figure 9-1 is used in Procedures 9-4 and 9-6. Calculate the value required for R_5 such that V_{ref} is approximately 2 V. Use the closest standard resistor value available in your parts kit. Show your work below.

$R_5 =$

Enter the value in Table 9-1.

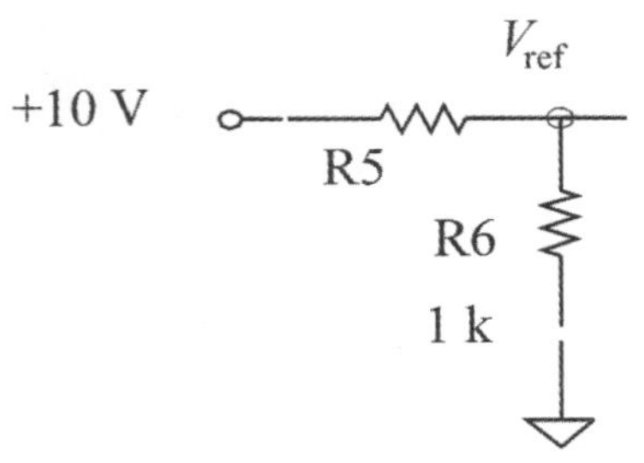

FIGURE 9-1 Resistor Voltage Design Circuit (*pre-lab*; part of Figures 9-3 and 9-5)

Using your standard value for R_5, determine the expected value for V_{ref}.

$V_{ref} =$ _____

Enter the value in Table 9-2.

1. Measure the resistors for Figure 9-1 and record them in Table 9-1. Calculate percent differences.

TABLE 9-1 Resistor Data Table for Figure 9-1

Resistor	Nominal Value	Measured Value	Percent Difference	Meet Specs?
R_5				☐ Yes ☐ No
R_6	I kΩ			☐ Yes ☐ No

2. Connect a 10-V supply to the circuit of Figure 9-1 and measure and record V_{ref} in Table 9-2. Keep this circuit for use later.

TABLE 9-2 Reference Voltage Circuit Design Measurement

Parameter	Expected Value	Measured Value	Percent Difference
V_{ref}			

Procedure 9-2

Noninverting Comparator without Feedback

SAMPLE CALCULATIONS (*pre-lab*)

Recall that the current into the + terminal of the op amp is very small. Therefore R_1, R_2 and R_{POT} essentially form a series circuit. Find:

$V_a =$

$V_b =$

$V_{out\ upper\ rail} =$

$$V_{\text{out lower rail}} =$$

$$V_{\text{trip b}\rightarrow\text{a}} =$$

$$V_{\text{trip a}\rightarrow\text{b}} =$$

1. Measure the resistors for Figure 9-2 and record them in Table 9-3. Calculate percent differences.

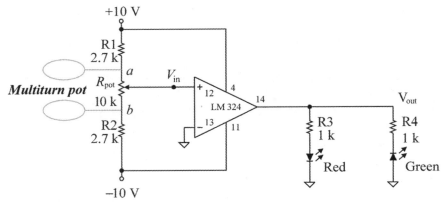

FIGURE 9-2 Op-Amp Comparator (pre-lab)

TABLE 9-3. Resistor Data Table for Figure 9-2

Resistor	Nominal Value	Measured Value	Percent Difference	Meet Specs?
R_{POT}	10 kΩ			☐ Yes ☐ No
R_1	2.7 kΩ			☐ Yes ☐ No
R_2	2.7 kΩ			☐ Yes ☐ No
R_3	1 kΩ			☐ Yes ☐ No
R_4	1 kΩ			☐ Yes ☐ No
R_i	100 kΩ			☐ Yes ☐ No
R_f	470 kΩ			☐ Yes ☐ No

2. Set up your + and −10-V supplies prior to connection to the circuit. Turn off the supplies before making the connections and then apply power.

3. Measure V_a and V_b, then adjust the multi-turn potentiometer until V_{in} is equal to V_a. This indicates that the potentiometer wiper is positioned at the top.

   Which LED is lit?   ☐ Red   ☐ Green

4. Measure and record the values in Table 9-4.

TABLE 9-4 Potentiometer Wiper Arm at Node **a** (top of POT) of Figure 9-2

Parameter	Expected Value	Measured Value	Percent Difference
$+E_{supply}$			
$-E_{supply}$			
V_{in} (same as V_a)			
V_{out}			
V_{R3}			
$V_{red\ LED}$			
V_{out} with scope*			

* Set the oscilloscope reference to the center of the screen so that both positive and negative voltages can be observed. Use the best scale possible to observe the voltage.

5. Now adjust the multi-turn potentiometer until V_{in} is equal to V_b. This indicates that the potentiometer wiper is positioned at the bottom.

   Which LED is lit? ☐ Red ☐ Green

6. Measure and record the values in Table 9-5.

TABLE 9-5 Potentiometer Wiper Arm at Node **b** (bottom of POT) of Figure 9-2

Parameter	Expected Value	Measured Value	Percent Difference
V_{in} (same as V_b)			
V_{out}			
V_{R4}			
$V_{green\ LED}$			
V_{out} with scope			

7. While observing the lamps, adjust the potentiometer wiper until the lamps switch. Record the V_{in} *trip points* in Table 9-6.

TABLE 9-6 Data to Determine Trip Voltage for the Circuit of Figure 9-2

Parameter	Expected Value	Measured Value	Percent Difference
$V_{in\ trip}$ (POT going from node **b** to **a**)			
$V_{in\ trip}$ (POT going from node **a** to **b**)			

Observations

1. Did the trip point differ significantly when going from **b** to **a** versus **a** to **b**? ☐ Yes ☐ No

2. How much headroom did your op amp have at the positive rail?

_____

3. How much headroom did your op amp have at the negative rail?

_____

Procedure 9-3

Open-Loop Noninverting Comparator with V_{ref}

SAMPLE CALCULATIONS

With your 2-V reference circuit added (as shown in Figure 9-3), find:

$V_{trip\ b \to a} =$

$V_{trip\ a \to b} =$

---

1. Turn off the + and −10-V supplies and add the 2-V reference circuit designed in Procedure 9-1 as shown in Figure 9-3.

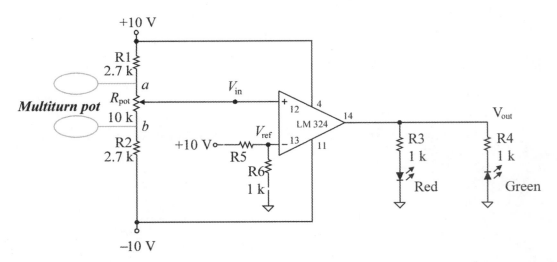

FIGURE 9-3 Noninverting Op-Amp Comparator with a Reference Voltage (Same circuit as Figure 9-2 with added reference circuit from Figure 9-1)

2. Apply power and recheck V_{ref}, V_a, and V_b. They should be the same as in the previous procedure.
3. Adjust the multi-turn potentiometer until V_{in} is equal to V_a.
   Which LED is lit? ☐ Red ☐ Green
4. Adjust the multi-turn potentiometer until V_{in} is equal to V_b.
   Which LED is lit? ☐ Red ☐ Green
5. While observing the lamps, adjust the potentiometer wiper until the lamps switch. Record the V_{in} trip points in Table 9-7.

TABLE 9-7 Data to Determine the Trip Voltage for the Circuit of Figure 9-3

Parameter	Expected Value	Measured Value	Percent Difference
V_{ref} (confirm prior measurement)		(STOP)	
$V_{in\ trip}$ (POT going from node **b** to **a**)		(STOP)	
$V_{in\ trip}$ (POT going from node **a** to **b**)		(STOP)	

(STOP) Instructor sign-off of the trip points in Table 9-7 _____

Observations ■

1. Did the trip point differ significantly when going from **b** to **a** versus **a** to **b**? ☐ Yes ☐ No
2. For the open-loop (without feedback) circuit configuration, what determines the trip point?

_____

3. What simple wiring change would modify this circuit to have a trip point at −2 V?

_____

Procedure 9-4

Noninverting Comparator with Positive Feedback

SAMPLE CALCULATIONS

The circuit of Figure 9-4 on page 120 uses positive feedback. With positive feedback V_{error} must equal 0 V to trip the op-amp output. Assume an op-amp headroom of 1.5 V for your calculations.

The wiper arm of Figure 9-4 is adjusted such that V_{in} is − 4 V and the op-amp output is slamming its lower rail. Find the expected values for:

$V_{Ri} =$

$V_{Rf} =$

$V_f =$

$V_{trip\ b \to a} =$

Now the wiper arm of Figure 9-4 is adjusted such that V_{in} is +4 V and the op-amp output is slamming its upper rail. Find the expected values for:

$V_{Ri} =$

$V_{Rf} =$

$V_f =$

$V_{trip\ a \to b} =$

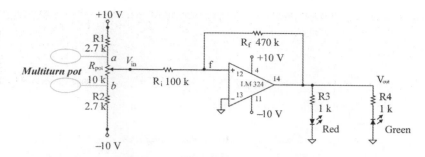

FIGURE 9-4   Noninverting, Positive Feedback Op-Amp Comparator

1. Turn off the + and −10-V supplies. Remove the reference circuit (R_5 and R_6) and add the feedback and input resistors (R_f and R_i) shown in Figure 9-4. Measure the feedback and input resistors (R_f and R_i) and record them in Table 9-8. Calculate percent differences.

TABLE 9-8   Resistor Data Table for Figure 9-4

Resistor	Nominal Value	Measured Value	Percent Difference	Meet Specs?
R_f	470 kΩ			☐ Yes ☐ No
R_i	100 kΩ			☐ Yes ☐ No

2. Apply power and recheck V_a and V_b. They should be the same as in previous procedures.
3. Adjust the multi-turn potentiometer until V_{in} is equal to −4 V. Which LED is lit?   ☐ Red ☐ Green
4. Measure and record the values in Table 9-9.

TABLE 9-9   Potentiometer Wiper Arm Set so V_{in} is −4-V Input

Parameter	Expected Value	Measured Value	Percent Difference
V_{in} (set to −4 V)			
V_{out}			
V_{Ri}			
V_{Rf}			
V_f			
V_{out} with scope			

5. While observing the lamps, adjust the potentiometer wiper until the lamps switch. Record the trip point in Table 9-10.

TABLE 9-10 Data to Determine Trip Voltage for the Circuit of Figure 9-4

Parameter	Expected Value	Measured Value	Percent Difference
$V_{in\ trip}$ (POT going from node **b** to **a**)			

6. Turn your potentiometer two turns in the opposite direction. Do the LEDs swap?

☐ Yes ☐ No

7. Adjust the multi-turn potentiometer until V_{in} is equal to +4 V. Which LED is lit?

☐ Red ☐ Green

8. Measure and record the values in Table 9-11.

TABLE 9-11 Potentiometer Wiper Arm Set so V_{in} Is 4-V Input

Parameter	Expected Value	Measured Value	Percent Difference
V_{in} (set to +4 V)			
V_{out}			
V_{Ri}			
V_{Rf}			
V_f			
V_{out} with scope			

9. While observing the lamps, adjust the potentiometer wiper until the lamps switch. Record the trip point in Table 9-12.

TABLE 9-12 Data to Determine Trip Voltage for the Circuit of Figure 9-4

Parameter	Expected Value	Measured Value	Percent Difference
$V_{in\ trip}$ (POT going from node **a** to **b**)			

Observations

1. Were the measured values with V_{in} at −4 V reasonably close to the expected values? Consider that the measured rail voltages may differ from the calculated values.

☐ Yes ☐ No

2. Is the measured input voltage required to trip the op amp to its upper rail reasonably close to your expected value?

☐ Yes ☐ No

3. Were the measured values with V_{in} at +4 V reasonably close to the expected values? Consider that the measured rail voltages may differ from the calculated values.

☐ Yes ☐ No

4. Is the measured input voltage required to trip the op amp to its lower rail reasonably close to your expected value?

☐ Yes ☐ No

5. What was the difference between the trip voltages?

Synthesis

Noninverting Comparator with Positive Feedback and a Reference Voltage

---

SAMPLE CALCULATIONS

The V_{ref} circuit designed in Procedure 9-1 is added to the circuit of Figure 9-4 to yield the circuit of Figure 9-5. Determine the new values for:

$V_{trip\ b\to a} =$

$V_{trip\ a\to b} =$

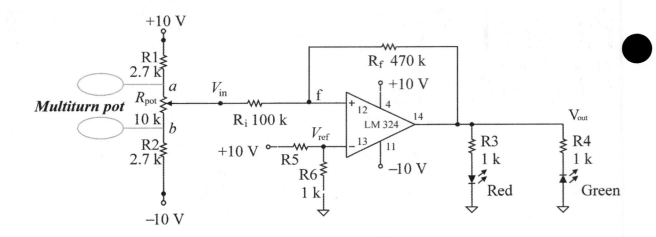

FIGURE 9-5 Noninverting, Positive Feedback Comparator with Reference (Same feedback circuit as Figure 9-4 with added reference circuit from Figure 9-1)

Turn off the supplies and add the reference circuit (R_5 and R_6) as shown in Figure 9-5. Recheck V_{ref}, V_a, and V_b. While observing the lamps, adjust the potentiometer wiper until the lamps switch and record the upper and lower trip points in Table 9-13.

TABLE 9-13 Data to Determine Trip Voltage for the Circuit of Figure 9-4

Parameter	Expected Value	Measured Value	Percent Difference
$V_{in\ trip}$ (POT going from node **b** to **a**)		(STOP)	
$V_{in\ trip}$ (POT going from node **a** to **b**)		(STOP)	

(STOP) Instructor sign-off of the trip points in Table 9-13 _____

Observations ▪

1. Is the measured input voltage required to trip the op amp to its upper rail reasonably close to your expected value?

   ☐ Yes   ☐ No

2. Is the measured input voltage required to trip the op amp to its lower rail reasonably close to your expected value?

   ☐ Yes   ☐ No

3. What was the difference between the trip voltages?

   _____

4. Was the difference between trip voltages the same as in Procedure 9-4?

   ☐ Yes   ☐ No

Parallel, Bridge, and DAC Circuits

Name: _____ Date: _____

Lab Section: _____ _____ Lab Instructor: _____
               day   time

Text Reference

DC/AC Circuits and Electronics: Principles and Applications
Chapter 10: Parallel Circuits

Materials Required

Triple power supply (2 @ 0–20 volts DC; 1 @ 5 volts DC)
Oscilloscope
4   2.2-kΩ resistors
2   10-kΩ resistors
2   1-kΩ resistors
1   each 100-kΩ, 1.8-kΩ, and 3.3-kΩ resistors
3   1.1-kΩ resistors (may be substituted with 1-kΩ + 100-Ω
1   10-kΩ multi-turn potentiometer
1   LM324 operational amplifier (op amp)

Introduction

In this exercise, you will:

- Analyse a parallel circuit and use Thévenin analysis on a bridge circuit
- Examine an R-2R ladder network and use an R-2R ladder network to build a digital-to-analog converter (DAC).

Pre-Lab Activity Checklist

☐ Perform pre-lab calculations for:

   ☐ Procedure 10-1 to find expected values in Tables 10-2 and 10-3

   ☐ Procedure 10-2 to find expected values in Tables 10-5 and 10-7

   ☐ Procedure 10-3 to find expected values in Tables 10-9 and 10-10.

☐ Build the circuits of Figures 10-2 and 10-5.

Performance Checklist

☐ Pre-lab completed?   🛑Instructor sign-off _____

☐ Demonstrate all measurements in Table 10-5.

☐ Demonstrate V_{out} for binary inputs 101, 110, 111 in Table 10-11.

Pre-Lab

Procedure 10 - 1

Parallel Circuit

SAMPLE CALCULATIONS

For the circuit of Figure 10-1, show your pre-lab calculations for:

$R_A =$

$R_B =$

$R_C =$

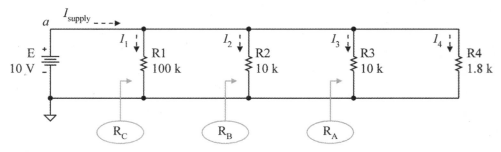

FIGURE 10-1 Parallel Circuit

Use Ohm's Law to find:

$I_1 = E/R_1 =$

$I_2 =$

$I_3 =$

$I_4 =$

I_{supply} with R$_1$ only =

Using KCL find:

I_{supply} with R$_1$ || R$_2$ =

I_{supply} with R$_1$ || R$_2$ || R$_3$ =

I_{supply} with R$_1$ || R$_2$ || R$_3$ || R$_4$ =

1. Measure the resistors for Figure 10-1 and record them in Table 10-1. Calculate percent differences.

TABLE 10-1 Resistor Data for Figure 10-1

Resistor	Nominal Value	Measured Value	Percent Difference	Meet Specs?
R$_1$	100 kΩ			☐ Yes ☐ No
R$_2$	10 kΩ			☐ Yes ☐ No
R$_3$	10 kΩ			☐ Yes ☐ No
R$_4$	1.8 kΩ			☐ Yes ☐ No

2. Use proper techniques to measure R_A ($R_3 \parallel R_4$). Repeat for R_B and R_C. Record the values in Table 10-2.

TABLE 10-2 Parallel Resistance Measurements Figure 10-1

Resistor	Expected Value	Measured Value	Percent Difference
R_A ($R_3 \parallel R_4$)			
R_B ($R_2 \parallel R_3 \parallel R_4$)			
R_C ($R_1 \parallel R_2 \parallel R_3 \parallel R_4$)			

3. Connect the 10-V supply and measure and record the values in Table 10-3.

TABLE 10-3 Currents in a Parallel Circuit

Parameter	Expected Value	Measured Value	Percent Difference
E			
V_a			
V_{R1}			
V_{R2}			
V_{R3}			
V_{R4}			
V_{R4} with CH1 on the scope			
I_{supply} with R_1			
I_{supply} with $R_1 \parallel R_2$			
I_{supply} with $R_1 \parallel R_2 \parallel R_3$			
I_{supply} with $R_1 \parallel R_2 \parallel R_3 \parallel R_4$			

Observations ■

1. How did adding resistors in parallel change the total resistance?

☐ Increased total resistance   ☐ Decreased total resistance

2. Did each resistor branch drop the same voltage?   ☐ Yes   ☐ No

3. How did adding a parallel branch change the total current supplied?

☐ Increased total current   ☐ Decreased total current

Procedure 10-2

Thévenin and Norton Analyses of a Bridge Circuit

SAMPLE CALCULATIONS

For the bridge circuit of Figure 10-2 find:

The open circuit voltage E_{TH}. Draw the bridge with an open load and find V_{ab}:

$E_{TH} =$

Enter the value in Table 10-5.

The short circuit current I_N. Draw the bridge with a short as a load and find I_{SC}.

$I_N = I_{SC} =$

Enter the value in Table 10-5.

The value of R_2 that will balance the bridge:

R_2 (balance) $=$

Enter the value in Table 10-5.

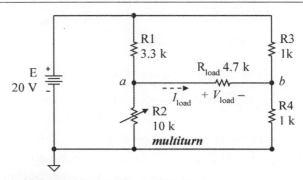

FIGURE 10-2 Bridge Circuit (*pre-lab*)

1. Measure the resistors for Figure 10-2 and record them in Table 10-4.

TABLE 10-4 Resistor Data Table for Figure 10-2

Resistor	Nominal Value	Measured Value	Percent Difference	Meet Specs?
R_1	3.3 kΩ			☐ Yes ☐ No
R_2	10 kΩ (POT) *at maximum*			☐ Yes ☐ No
R_3	1 kΩ			☐ Yes ☐ No
R_4	1 kΩ			☐ Yes ☐ No
R_{load}	4.7 kΩ			☐ Yes ☐ No

2. Set the multi-turn potentiometer to maximum value.
3. Connect the 20-V supply and measure and record the values in Table 10-5.

TABLE 10-5 Thévenin and Norton Model Values

Parameter	Expected Value	Measured Value	Percent Difference
E with DMM	20 V	🛑	
E_{TH} (open circuit voltage) with DMM		🛑	
I_N (short circuit current) with DMM		🛑	
V_a (open circuit) with scope CH1		🛑	
V_b (open circuit) with scope CH2		🛑	
V_{ab} (open circuit) with scope CH1-CH2		🛑	

🛑 Instructor sign-off of measured values in Table 10-5 _____

4. Using your measured values for E_{TH} and I_{SC}, calculate R_{TH} and draw the Thévenin model for terminals a and b of the bridge circuit in Figure 10-3.

FIGURE 10-3 Thévenin Model of Terminals a and b of Bridge Circuit

Calculations Using measured values for E_{TH} and I_{SC} find:

$R_{TH} =$

5. Using your Thévenin model (Figure 10-3), find the expected values for Table 10-6. Then, with the 4.7-kΩ load attached to the bridge circuit, measure the actual load current and voltage. Record the values in Table 10-6.

TABLE 10-6 Loaded Bridge

Parameter	Expected Value from Model	Measured Value	Percent Difference
V_L			
I_L			

Calculations Using your measured values, draw the Thévenin model in Figure 10-3. Then, find the current and voltage through the 4.7-kΩ load.

$I_L =$

$V_L =$

6. Calculate the potentiometer resistance required to "balance" the bridge. This is the value that will make the load voltage 0 V. (In other words, $E_{TH} = 0$ V; therefore, with the load removed $V_a = V_b$.)

Calculations Find the potentiometer resistance required to balance the bridge.

$R_{POT\_Balanced} =$

7. Adjust the 10-kΩ potentiometer (R_2) until the bridge is balanced. Remove the potentiometer and record the resistance in Table 10-7.

TABLE 10-7 Value of R_2 Needed to Balance the Bridge

Parameter	Expected Value	Measured Value	Percent Difference
R_2 (balanced bridge)			

Observations

1. Were the measured values of E_{TH} and I_{SC} close to the expected values? ☐ Yes ☐ No
2. Did the Thévenin model accurately predict your circuit load current and load voltage?
   ☐ Yes ☐ No
3. Was the measured value of R_2 to balance the bridge close to the calculated resistance value to balance the bridge?
   ☐ Yes ☐ No

Procedure 10-3

R-2R Ladder Circuit

The R-2R ladder circuit shown in Figure 10-4 is a novel method of creating constant current paths. Because of the R-2R resistor relationship, currents in each adjacent branches have a 2:1 ratio. We are using a three-branch ladder, so ideally the branch currents will have a 4:2:1 ratio.

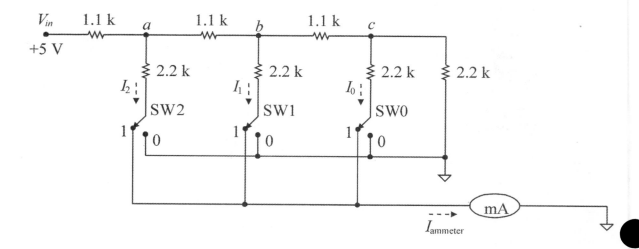

FIGURE 10-4 R-2R Ladder Circuit Shown with All Switches Connected to Current Meter

Notice that, regardless of switch position, as one looks to the right of point c the resistance is 1.1 kΩ. Similarly, as one looks to the right of points a and b, the resistance is also always 1.1 kΩ. The switches allow currents to either bypass (0 position) or pass through (1 position) the ammeter.

SAMPLE CALCULATIONS

Draw the ladder network and use resistor reduction to find R_{total}.

Ladder circuit $R_{total} =$

Using VDR find

$V_a =$

$V_b =$

$V_c =$

Using Ohm's Law find

$I_0 = I_{ammeter} (001) =$

$I_1 = I_{ammeter} (010) =$

$I_2 = I_{ammeter} (100) =$

Using KCL find

$I_{ammeter} (011) = I_1 + I_0 =$

$I_{ammeter} (101) =$

$I_{ammeter} (110) =$

$I_{ammeter} (111) =$

1. Measure the resistors for Figures 10-4 and 10-5 (used later) and record them in Table 10-8.

TABLE 10-8 Resistor Data Table for Figures 10-4 and 10-5 (measure Rs left to right)

Nominal Value	Measured Value	Percent Difference	Meet Specs?
1.1 kΩ			☐ Yes ☐ No
1.1 kΩ			☐ Yes ☐ No
1.1 kΩ			☐ Yes ☐ No
2.2 kΩ			☐ Yes ☐ No
2.2 kΩ			☐ Yes ☐ No
2.2 kΩ			☐ Yes ☐ No
2.2 kΩ			☐ Yes ☐ No
1 kΩ			☐ Yes ☐ No

Note: If 1.1-kΩ resistors not available, use a 1-kΩ in series with a 100-Ω resistor. The 1-kΩ is for use in the op-amp circuit of Figure 10-5.

2. Remove the connection between the R-2R ladder and the op amp of the circuit built in pre-lab. Connect the ammeter appropriately.
3. Set all switches to the 0 position and measure and record the R_{total} values in Table 10-9.
4. Connect the 5-V supply and measure and record the remaining values in Table 10-9. Calculate the ratios of the measured node voltage to the supply voltage. (If you have three adjustable supplies available, set V_{in} to 7 V.)

TABLE 10-9 Total Resistance and Voltages for Figure 10-4

Parameter	Expected Value	Measured Value	Percent Difference	Ratio of $V_{measured}$ to V_{in}
R_{total} switches position 0				
V_{in}				1 :1
V_a				:1
V_b				:1
V_c				:1

5. Carefully connect and disconnect switches to complete the ammeter readings for Table 10-10. Calculate percent differences and the ratios of the measured current to the current in the 001 position. (*Use this circuit for the Synthesis Procedure.*)

TABLE 10-10 Ammeter Values for all Possible Switch Positions of Figure 10-4

Switches SW2-SW1-SW0	$I_{ammeter}$ Expected Value	$I_{ammeter}$ Measured Value	Percent Difference	Ratio of $I_{measured\ 001}$ to $I_{measured}$
000	0 mA			I: 0
001				I: I
010				I:
011				I:
100				I:
101				I:
110				I:
111				I:

Observations ▪

1. Was the measured value of R_{total} close to the expected values? ☐ Yes ☐ No
2. Was the relationship of V_a, V_b, and V_c approximately 4:2:1 (binary)? ☐ Yes ☐ No
3. Were the branch currents binary in nature (that is, $I_2 = 2I_1$, and $I_1 = 2I_0$)? ☐ Yes ☐ No
4. Was the total load current with all of the switches closed equal to the sum of the individual current contributions of I_0, I_1, and I_2? Is KCL satisfied? ☐ Yes ☐ No

Synthesis
Digital-to-Analog Converter with R-2R Circuit

Turn off the power supply and reconnect the switch-1 positions to the input of the op amp as shown in Figure 10-5. Your circuit should look like Figure 10-6. Set the op-amp supply voltages prior to connecting them and then turn off the supply. Apply power and complete the measurements in Table 10-11. If you have three adjustable supplies, set V_{in} to 7 V.

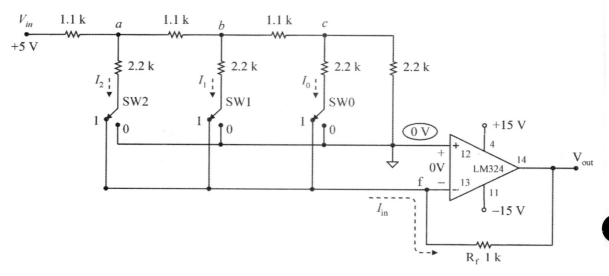

FIGURE I0-5 Digital-to-Analog (DAC) Circuit with R-2R Ladder Circuit (*pre-lab*)

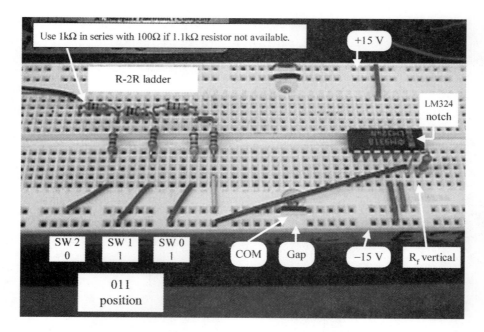

FIGURE I0-6 Board Layout for R-2R Digital-to-Analog (DAC) Circuit (*pre-lab*)

TABLE 10-11 Circuit Values for Figure 10-4

Switch states SW 2-1-0	*Measured I_{in}	*Expected V_{out}	Measured V_{out}	**Ratio of measured V_{out} / V_{out} (001)
0 0 0	0 mA			
0 0 1				1 : 1
0 1 0				
0 1 1				
1 0 0				
1 0 1			(STOP)	
1 1 0			(STOP)	
1 1 1			(STOP)	

*Use the *measured* values from Table 10-10 for I_{in}, and calculate the expected V_{out} based on that measurement.
** Divide each measured value of V_{out} by the measured value of V_{out} in position 001.

(STOP) Instructor sign-off of the 101, 110, and 111 readings _____

SAMPLE CALCULATION
_____

Using the *measured* values for I_{in} from Table 10-10 find:

V_{out} with all switches closed =

Observations ■

1. Does digital input of 101 translate to an analog value of five times the 001 input?

   ☐ Yes   ☐ No

2. Does the circuit give an analog output proportional to the digital input?

   ☐ Yes   ☐ No

3. Summarize the operation of the digital-to-analog converter (DAC) circuit. Use complete sentences and proper punctuation.

_____

_____

_____

_____

_____

_____

Alternating Waveforms

Name: _____   Date: _____

Lab Section: _____ _____   Lab Instructor: _____
                      day   time

Text Reference

DC/AC Circuits and Electronics: Principles and Applications
Chapter 12: Waveforms

Materials Required

Triple power supply (2 @ 0–20 volts DC; 1 @ 5 volts DC)

Oscilloscope

Function generator 1 Hz–100 kHz (sine, triangle, square waveforms; DC offset)

Audio headset

Piezo sound transducer (buzzer)

Microphone (optional)

3 10-kΩ resistors

1 LM324 operational amplifier (op amp)

Introduction

In this exercise, you will:

- Examine AC waveforms including sinusoidal, triangular, and square waveforms
- Compare the RMS voltages of sinusoidal, triangular, and square waveforms
- Observe the effects of amplitude and frequency variations using an oscilloscope, an audio headset, and a transducer
- Add DC offset to an AC signal
- Use a summing inverter amplifier to add a DC offset to an AC signal.

Pre-Lab Activity Checklist

☐ Perform pre-lab calculations for:

　☐ Procedure 11-1 to find expected values in Table 11-1

　☐ Procedure 11-2 to find expected values in Table 11-2

　☐ Expected values in Table 11-3

　☐ Procedure 11-4 to find expected values in Tables 11-4 and 11-5

　☐ Procedure 11-5 to find expected values in Table 11-6.

☐ Build the circuit of Figure 11-7.

Performance Checklist

☐ Pre-lab completed?　　　　　　　　　🛑 Instructor sign-off _____

☐ Demonstrate measurements in Table 11-1 and Figure 11-3.

☐ Demonstrate 60-Hz measurement in Table 11-6.

☐ Demonstrate oscilloscope displays reproduced in Figures 11-8 and 11-9.

Procedure 11-1

Observation of an 8-V$_{pp}$, 1-kHz Sinusoidal Signal

SAMPLE CALCULATIONS

Find the RMS value of the 8-V$_{pp}$, 1-kHz AC signal:

$$V_{rms} = \frac{V_{pp}}{2\sqrt{2}} =$$

Here is how to calculate the best V/DIV and SEC/DIV selections to observe an 8-V$_{pp}$, 1-kHz sine wave on the oscilloscope:

V/DIV　　　$\dfrac{V_{pp}}{8\ \text{DIV}} = \dfrac{8\ \text{V}}{8\ \text{DIV}} = 1\ \text{V/DIV}$　　　　Exact control setting*

SEC/DIV　　$T = 1/f = 1/1\ \text{kHz} = 1\ \text{ms}$　　　　Period of waveform

　　　　　　$\dfrac{T}{10\ \text{DIV}} = \dfrac{1\ \text{ms}}{10\ \text{DIV}} = 0.1\ \text{ms/DIV}$　　　　Exact control setting*

*Normally the V/DIV or SEC/DIV calculated is not an exact setting; in that case one should round up to the next highest control setting.

1. Set up the function generator as shown in Figure 11-1. Your generator may differ; see your instructor for assistance.

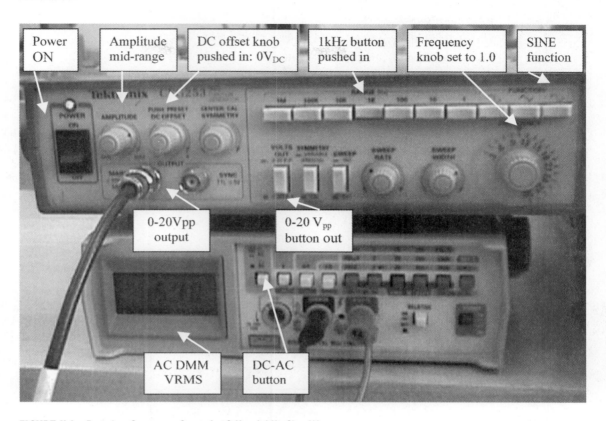

FIGURE II-I   Function Generator Setup for 8-V$_{pp}$, I-kHz Sine Wave

2. Connect the oscilloscope to the function generator output and set up your oscilloscope to properly measure this signal on CH1 (Figure 11-2).

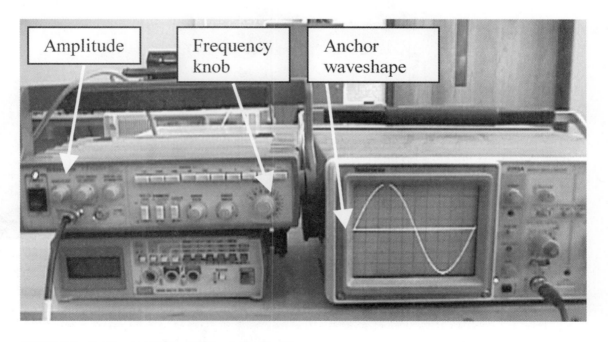

FIGURE II-2   Oscilloscope Display of 8-V$_{pp}$, I-kHz Sine Wave

Use the following settings:

☐ GND line centered

☐ AC mode setting

☐ 1 V/DIV to observe 8 V_{pp}

☐ 0.1 mSEC/DIV to observe 1 mS period

☐ Adjust *level* and *position* to anchor sine wave at 0-V line

3. Adjust the function generator *amplitude* and *frequency* knob until the desired signal is achieved on the oscilloscope (8 vertical DIVs and 10 horizontal DIVs).

4. Sketch the waveform observed on the oscilloscope in Figure 11-3. Be sure to fill in the V/DIV, SEC/DIV, V(___), and t(___). Show the 0-V line (your ground line placement). Show scale values (e.g., 4 V, 0 V, +4 V, and 0 ms, 0.5 ms, and 1 ms).

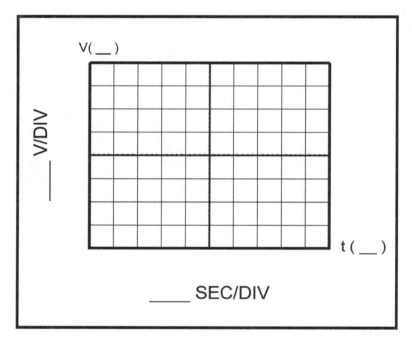

FIGURE II-3 Sketch the Observed Signal: of 8 V_{pp}, 0 V_{DC}, I kHz

5. Measure the DMM AC and DC values in Table 11-1.

TABLE II-I Scope and DMM Comparison of RMS Values		
Variable	**Expected Value**	**Measured Value**
Scope V_{pp}	8 V_{pp}	
DMM AC (RMS)		STOP
DMM DC		STOP

STOP Instructor sign-off of measured values in Figure 11-3 and Table 11-1 _____

Observations ■

1. Did the scope display and DMM measurements show equivalent values?   ☐ Yes   ☐ No
2. Which is the most appropriate instrument to measure peak and peak-to-peak voltages?

☐ Oscilloscope   ☐ DMM

3. Which is the most appropriate instrument to measure RMS and DC voltage?

☐ Oscilloscope   ☐ DMM

Procedure 11-2

Observation of a 2-V$_{pp}$, 1-kHz Sinusoidal Waveform

SAMPLE CALCULATIONS

Find the RMS value of the 2-V$_{pp}$, 1-kHz AC signal:

$$V_{rms} = \frac{V_{pp}}{2\sqrt{2}} =$$

Here is how to calculate the best V/DIV and SEC/DIV selections to observe a 2-V$_{pp}$, 1-kHz sine wave on the oscilloscope:

V/DIV $\dfrac{V_{pp}}{8 \text{ DIV}} = \dfrac{2 \text{ V}}{8 \text{ DIV}} = 0.25 \text{ V/DIV} \rightarrow 0.5 \text{ V/DIV}$ (next highest setting)

When displayed using 0.5-V/DIV setting:

$$\# \text{ DIV displayed} = \frac{2 \text{ V}}{0.5 \text{V/DIV}} = 4 \text{ DIV} \rightarrow 4 \text{ divisions amplitude}$$

---

1. Adjust the function generator so that the DMM reads the RMS value of 2 V$_{pp}$ calculated in the pre-lab. Record the measured and expected values in Table 11-2.

TABLE 11-2 Scope and DMM Comparison of Values

Variable	Expected Value	Measured Value	Percent Difference
DMM AC (RMS)			
Scope V_{pp} reading			

2. Display the signal on the oscilloscope in the AC coupling mode with the ground line centered. Adjust your V/DIV to the best setting found above (0.5 V/DIV), then measure the peak-to-peak voltage and record the value in Table 11-2.

3. Maintain these setting for use in Procedure 11-3.

Observation ▪

1. Did your scope display the 2 V_{pp} expected?   ☐ Yes ☐ No

Procedure 11-3
Adding DC Offset to an AC Waveform

1. Locate the "DC Offset" control of your function generator. Often you will pull out a knob to engage the DC offset.

2. With the oscilloscope in the AC coupling mode, set the DMM in the DC position and adjust the generator DC offset to +6 V_{dc}.
   Was there any difference in the oscilloscope display?   ☐ Yes ☐ No

   If so, describe it. _____

   _____

   ◨ **Note:** Using AC coupling removes the DC portion of the signal.

3. Now switch to DC coupling. This includes the DC portion of the signal as well as the AC.
   Was there any difference in the oscilloscope display?   ☐ Yes ☐ No

   If so, describe it. _____

   _____

4. Change the V/DIV setting to 1 V/DIV. Set the ground line position to the bottom of the scale, and then return to DC coupling. The entire waveform should now be visible. Sketch the waveform on Figure 11-4 including all scales, V/DIV and SEC/DIV.

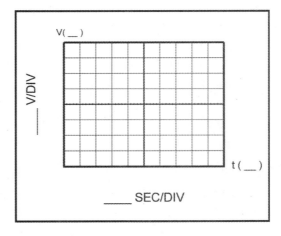

FIGURE 11-4 Sketch the Observed Sine Wave: 2 V_{pp}, V_{DC}, 1 kHz

5. Some DMMs read only the AC portion of a waveform when in the AC mode. Others read the RMS value of the AC+DC. Still others may perform a calculation based upon the AC+DC and display a reading that may be meaningless.

6. Using your oscilloscope and DMM, record the measurements in Table 11-3.

TABLE II-3 Scope and DMM Comparison of RMS Values for the **Sine** Wave

Variable	Expected Value	Measured Value
Scope V_{max}		
Scope V_{min}		
Scope V_{pp}		
Scope V_p (V_{pp} /2)		
Scope V_{DC}		
DMM AC (RMS)		
DMM DC		

7. Switch to AC coupling and slowly rotate the DC offset control while observing the oscilloscope.

8. Switch to DC coupling and again slowly rotate the DC offset control while observing the oscilloscope.

Observations ▪

1. Did the scope display and DMM measurements show equivalent values? ☐ Yes ☐ No

2. To calculate the total RMS of a waveform, take the square root of the sum of the squares of the individual components. Using ideal values:

$$V_{rms\ total} = (V_{DC}^2 + V_{1\ kHz}^2)^{\frac{1}{2}} = (6^2 + 0.707^2)^{\frac{1}{2}} = 36.5^{\frac{1}{2}} = \boxed{6.042\ V_{rms}}$$

Using the *measured* AC RMS without offset (from Procedure 11-2) and DC DMM values, calculate your total RMS voltage.

3. Based upon your DMM reading, does your meter read the total RMS in the AC mode for a sinusoidal waveform?

☐ Yes ☐ No

4. Describe the effect of rotating the DC offset control while using AC coupling.

_____

_____

5. Describe the effect of rotating the DC offset control while using DC coupling.

_____

_____

Procedure 11-4

Sinusoidal, Triangular, and Square Waveforms

SAMPLE CALCULATIONS

Find the corresponding RMS voltages for a 2-V_{pp} waveform (0-V_{DC} offset).

Sine wave

$$V_{\text{true rms}} = V_{\text{rms}} = \frac{V_{pp}}{2\sqrt{2}} =$$

Triangle wave

$$V_{\text{true rms}} = V_{\text{rms}} = \frac{V_{pp}}{2\sqrt{3}} =$$

Square wave

$$V_{\text{true rms}} = V_{\text{rms}} = \frac{V_{pp}}{2}$$

---

1. Remove the DC offset from your AC signal while maintaining 2 V_{pp}. Set the V/DIV to 0.5 V/DIV. Center the ground position on the oscilloscope, then return to DC coupling.

2. Switch the function generator to display a triangular waveform. If the peak-to-peak voltage is not still 2 V_{pp} as viewed on the scope, readjust the function generator's *amplitude*.

3. Now sketch the waveform in Figure 11-5. Measure the peak-to-peak voltage using the scope and record it in Table 11-4.

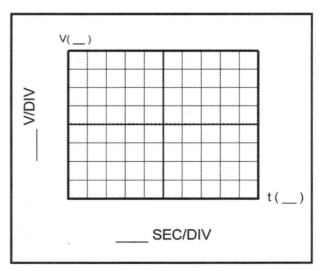

FIGURE 11-5 Sketch the Observed **Triangle** Wave: 2 V$_{pp}$, 1 kHz (*Fill in the appropriate values on your Figure.*)

TABLE 11-4 Scope and DMM Comparison for the **Triangle** Wave

Variable	Expected Value	Measured Value
Scope V_{pp}	2 V_{pp}	
DMM AC (RMS)		

4. Some more expensive DMMs read the true RMS value of a waveform. Most perform a calculation based upon the peak value of the applied signal. The meter then scales the display number based upon the peak of a sinusoidal waveform. For signals other than a sinusoid, the reading may be meaningless.

5. Measure the RMS voltage of the triangle wave using the DMM and record the results in Table 11-4.

6. Change the signal to a square wave and complete Figure 11-6 and Table 11-5.

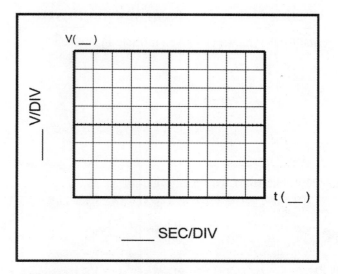

FIGURE 11-6 Sketch the Observed **Square** Wave: 2 V_{pp}, 1 kHz

TABLE 11-5 Scope and DMM Comparison for the **Square** Wave

Variable	Expected Value	Measured Value
Scope V_{pp}	2 V_{pp}	
DMM AC (RMS)		

Observations ■

1. Based upon your DMM reading, does your meter read the true RMS in the AC mode for a triangular waveform? ☐ Yes ☐ No

2. If not, calculate the correction factor for a triangular waveform:

   Triangle wave DMM correction factor = $V_{\text{true rms}} / V_{\text{DMM}}$ = _____ = _____

3. Based upon your DMM reading, does your meter read the true RMS in the AC mode for a square wave-form?

☐ Yes ☐ No

4. If not, calculate the correction factor for a square waveform:

Square wave DMM correction factor = $V_{\text{true rms}} / V_{\text{DMM}}$ = _____ = _____

Procedure 11-5

Generator Frequency Accuracy

SAMPLE CALCULATIONS

Find the period for the frequencies given in Table 11-6. Show the calculation for 60 Hz.

$T_{60\text{Hz}} =$

TABLE 11-6. Function Generator Frequency Measurements

Use Function Generator Controls to Set Up the Approximate Frequency	Expected Period T (pre-lab)	Scope Measured Period (T)	Scope Measured Frequency (f)	Percent Difference
60 Hz		(STOP)	(STOP)	
1 kHz				
20 kHz				
85 kHz				
330 kHz				

(STOP) Instructor sign-off of 60-Hz oscilloscope display _____

1. Leave the amplitude set at 2 V_{pp} with 0.5 V/DIV sensitivity on CH1 of the scope. Change the frequency of the generator to 60 Hz using only the function generator controls.

2. Use the oscilloscope to measure the period and then calculate the frequency. It will most likely **not** be 60 Hz. Record the period and frequency measurements in Table 11-6.

3. Repeat this procedure using the function generator controls to set the remaining frequencies in Table 11-6.

Observations

1. Based upon your measurements, are the function generator frequency controls accurate?

   ☐ Yes ☐ No

2. Which instrument should be used to *accurately* set a signal's frequency (period)?

   ☐ Function generator ☐ Oscilloscope

Procedure 11-6

Audio Headset

1. Set the function generator amplitude to minimum (~ 0 V$_{pp}$) and the frequency to ~ 1 kHz (sinusoidal wave-form). Connect the generator to both an oscilloscope and an audio headset.

2. Slowly increase the amplitude of the generator signal to a comfortable sound level. Then vary the function generator *amplitude* knob and note the effect on the sound and on the oscilloscope display.

3. Reset the volume to a comfortable level, then vary the function generator *frequency* knob and note the effect on the sound and on the oscilloscope display.

4. Start with a frequency of 1 Hz and slowly increase the signal generator frequency until you can just hear the tone (low frequency cutoff). Use the measured period of this signal (using the oscilloscope) and calculate the frequency. Record these values.

   $T_{low} =$ _____ $f_{low} =$ _____

5. Start with a frequency of 10-kHz and slowly increase the signal generator frequency until you can no longer hear the tone (high frequency cutoff). Use the measured period of this signal (using the oscillo-scope) and calculate the frequency. Record these values.

   $T_{high} =$ _____ $f_{high} =$ _____

6. Start with a frequency of 1 kHz and change the waveform to triangular. Note the difference in tone.

7. Start with a frequency of 1 kHz and change the waveform to square. Note the difference in tone.

8. Disconnect the headset.

Observations

1. Is amplitude related to volume or pitch?

   ☐ Volume ☐ Pitch

2. Is frequency related to volume or pitch?

   ☐ Volume ☐ Pitch

3. Does a radio volume control change the amplitude or the frequency of the audio signal?

   ☐ Amplitude ☐ Frequency

4. Calculate the bandwidth of the headset as tested with your ears (by listening). (It may be limited by your ears!)

   Bandwidth: BW $= f_{high} - f_{low} =$ _____ $=$ _____

Procedure 11-7

Using a Sound Transducer as a Speaker

1. Set the function generator amplitude to minimum (~0 V_{pp}) and the frequency to ~1 kHz (sinusoidal wave-form). Connect the generator to both an oscilloscope and a sound transducer (buzzer).
2. Following the procedure previously used, measure the upper and lower frequencies that you can hear with the transducer. Do this quickly to avoid annoying others. Record these values.

    $T_{low} =$ _____ $f_{low} =$ _____

    $T_{high} =$ _____ $f_{high} =$ _____

3. Increase the frequency and note the frequency at which sound is the most intense (piercing). This is the resonant frequency of the transducer. Quickly record the period and frequency and reduce the amplitude to minimum.

    $T_{resonant} =$ _____ $f_{resonant} =$ _____

4. Using a frequency of 1 kHz, change the waveform to square and triangular and note the difference in tone.

Observation ◼

1. Calculate the bandwidth of the sound transducers as tested with your ears (by listening). (Again, it may be limited by your ears!)

    Bandwidth: BW $= f_{high} - f_{low} =$ _____ $=$ _____

Procedure 11-8

Using a Sound Transducer as a Microphone

1. Disconnect the generator, but leave the oscilloscope and a sound transducer connected.
2. Observe the oscilloscope while tapping, whistling, and talking into the transducer. You may need to adjust both V/DIV and SEC/DIV settings.
3. If a microphone is available, compare the output amplitude of the microphone to that of the sound transducer.

Observations ◼

1. A transducer converts one form of energy into another.
   This transducer converts _____ energy to or from _____ energy.
2. Is loudness associated with the signal's amplitude or frequency?

    ☐ Amplitude ☐ Frequency

3. Is pitch associated with the signal's amplitude or frequency?

    ☐ Amplitude ☐ Frequency

4. Does increasing the signal's period increase or decrease pitch?

    ☐ Increase ☐ Decrease

Synthesis

Adding DC Offset to an AC Signal

Measure and record the resistor values in Table 11-7. Adjust power supplies before connecting to the circuit of Figure 11-7 and turn the supply off before making circuit connections. Use the DMM to measure and record the DC and RMS values of Table 11-8 on page 152.

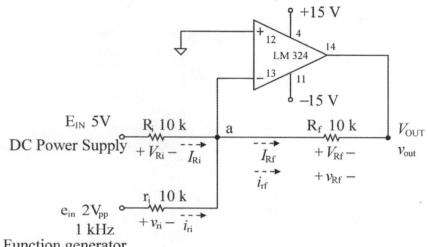

FIGURE II-7 Inverting Summing Amplifier: Adding a DC Offset to an AC Signal (*pre-lab*)

TABLE II-7 Resistor Data Table for Figure II-7

Resistor	Nominal Value	Measured Value	Percent Difference	Meet Specs?
R_i	10 kΩ			☐ Yes ☐ No
r_i	10 kΩ			☐ Yes ☐ No
R_f	10 kΩ			☐ Yes ☐ No

Use the oscilloscope and observe/sketch the input AC voltage and the output AC voltage on Figure 11-8 on page 153. For comparison, use the dual-trace mode with the settings below to display both the input and output signals at the same time to compare them.

Display   BOTH   ALT   Triggered INT on CH1 signal

CH 1 e_{in} GND line centered on top half of screen   AC coupling

CH 2 v_{out} GND line centered on bottom half of screen   AC coupling

               Not INVerted

TABLE II-8 Summing Amplifier

Parameter		Expected Value	Measured Value	Percent Difference
$+ E_{supply}$	DC	15 V		
$- E_{supply}$	DC	$- 15$ V		
E_{in}	DC	$+5$ V		
V_a	DC	$\cong 0$ V		
V_{Ri}	DC	5 V		
I_{Ri}	DC	500 μA		
V_{Rf}	DC	5 V		
I_{Rf}	DC	500 μA		
V_{out}	DC	$- 5$ V		
T		I ms		
f		I kHz		
e_{in}	AC	0.707 V		
v_a	AC	$\cong 0$ V		
v_{ri}	AC	0.707 V		
i_{ri}	AC	71 μA		
v_{Rf}	AC	0.707 V		
i_{Rf}	AC	71 μA		
V_{out}	AC	0.707 V		
$V_{out\ pp}$	SCOPE	2 V		
$V_{out\ peak}$	SCOPE	I V		
$V_{out\ max}$	SCOPE	$- 4$ V		
$V_{out\ min}$	SCOPE	$- 6$ V		

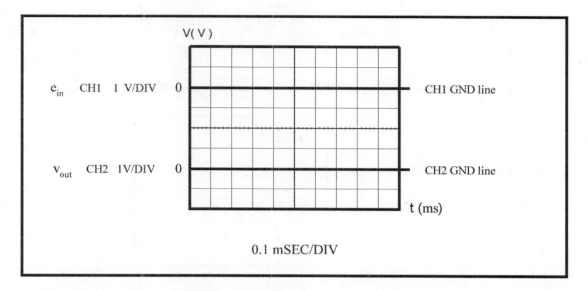

FIGURE 11-8   Dual-Trace CH1 and CH2, AC MODE (triggered INTernally on CH1)

Use the oscilloscope and observe/sketch the total output voltage waveform (AC with DC offset) on Figure 11-9. Use the single-trace mode with the following settings:

Display   CH2   ALT   Triggered INT on CH1 signal

CH 2 v_{out} GND top line   DC coupling (total signal)   Not INVerted

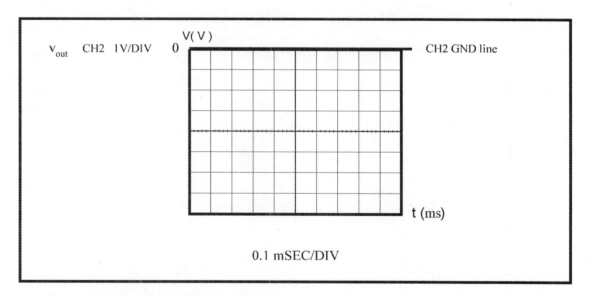

FIGURE 11-9   [STOP] CH2, DC MODE (total output signal).
(*Label CH2 values: maximum, minimum, and DC values.*)

(STOP)   Instructor sign-off of Figure 11-8 and 11-9 oscilloscope displays

Increase the amplitude of the AC signal (function generator) until the signal "clips" (top or bottom flattens out). Adjust the V/DIV setting as needed. Sketch the total output voltage waveform (AC with DC offset) on Figure 11-10.

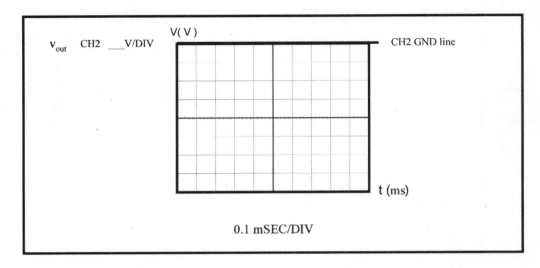

FIGURE 11-10 CH2, DC MODE (total output signal) with Clipping.
Label CH2 values: maximum, minimum, and DC values.)

AC and DC Electronic Circuits

Name: _____   Date: _____

Lab Section: _____ _____   Lab Instructor: _____
              day   time

Text Reference ◼

DC/AC Circuits and Electronics: Principles and Applications
Chapter 13: Capacitance and Reactance

Materials Required ◼

Triple power supply (2 @ 0–20 volts DC; 1 @ 5 volts DC)

Oscilloscope

Capacitance meter

Function generator 1 Hz–100 kHz (sine, triangle, square waveforms; DC offset)

1   1N4001 diode

1   2N3904 transistor

1   each 470-Ω, 820-Ω, 1-kΩ, 3.3-kΩ, 10-kΩ, 100-kΩ resistor

1   resistor determined by student

1   100-µF capacitor

1   LM324 operational amplifier (op amp)

Introduction ◼

In this exercise, you will:

- Examine a half-wave rectifier circuit with and without a filtered output
- Observe circuit gain using a bipolar junction transistor (BJT)
- Use an operational amplifier as a buffer amplifier to minimize circuit loss
- Design an operational-amplifier circuit to have a specific voltage gain.

Pre-Lab Activity Checklist

☐ Perform pre-lab calculations for:

    ☐ Procedure 12-1 to find expected values in Table 12-2

    ☐ Procedure 12-2 to find expected values in Tables 12-5 and 12-6

    ☐ Synthesis procedure to find expected values in Tables 12-11 and 12-12.

☐ Find the expected values in Tables 12-9 and 12-10.

☐ Build the circuits of Figure 12-5 and Figure 12-9(b).

Performance Checklist

☐ Pre-lab completed?   🛑 Instructor sign-off _____

☐ Show the measurements in Tables 12-2 and 12-3 and in Figures 12-2, 12-3, and 12-4.

☐ Demonstrate all measurements in Table 12-6.

☐ Show measurements in Table 12-12.

Procedure 12-1

Half Wave Rectifier

SAMPLE CALCULATIONS

Calculate period and the peak output voltage, V_p, of a half-wave rectified sinusoid. This will be the peak output voltage ($V_{out\text{-}p}$) for the circuits of both Figures 12-1(a) and 12-1(b).

$T =$

$V_{p\ \text{without or with capacitor}} =$

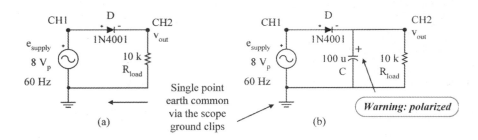

FIGURE 12-1   Half-Wave Rectifier (**a**) without Filter Capacitor and (**b**) with Filter Capacitor

1. Measure the values of R_{load} and C and record in Table 12-1.

Table 12-1 Resistor Data Table for Figure 12-1

Resistor	Nominal Value	Measured Value	Percent Difference
R	10 kΩ		
C	100 μF		

2. Set up the function generator using CH1 of the oscilloscope to accurately set up the generator signal for Figure 12-1. Record actual period and voltage in Table 12-2. Also, measure its DC voltage (it should be very close to 0 V since the DC OFFSET of the generator should be off) and record it in Table 12-2.

Table 12-2 Data Table for Figure 12-1

Parameter		Expected Value	Measured Value	Percent Difference
T of e_{supply}				
f of e_{supply}		60 Hz		
$e_{supply\ p}$	peak voltage	8 V		
$e_{supply\ dc}$	DMM			
$v_{out\ p}$	without capacitor		STOP	
$v_{out\ dc}$	DMM	2.3 V	STOP	
$v_{out\ p}$	with capacitor		STOP	
$v_{out\ dc}$	DMM	7.3 V	STOP	

3. Connect the circuit of Figure 12-1. Connect CH1 and CH2 of the scope. **Remember: All oscilloscope ground leads must be connected to common—a single point in the circuit.** Sketch the waveforms in Figure 12-2 making CH1 a *dashed line* and CH2 a *solid line*. Clearly label CH1 and CH2 waveforms on your sketch. Measure the peak voltages of both waveforms and record them in Table 12-2. Measure the DC output voltage with the DMM and record it in Table 12-2.

Compare Input and Output Voltages (with no capacitor filter)

BOTH	CHOP or ALTernate (*best*)	triggered INTernal on CH1 signal	
CH 1: e_{supply}	GND line centered	DC Coupling	
CH 2: v_{out}	GND line centered	DC Coupling	Not INVerted

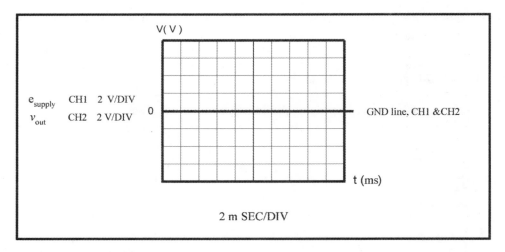

FIGURE 12-2   Half-Wave Rectifier *without* Filter Capacitor

4. Turn off the generator and connect the filter capacitor to the circuit.

 WARNING: Be very CAREFUL, this is a polarized capacitor and must be connected properly or it could explode. As always you must wear safety glasses in lab.

5. Reapply power and sketch the waveforms for the circuit including the capacitor in Figure 12-3. Sketch CH1 as a *dashed line* and CH2 as a *solid line*. Measure the peak voltages of the output voltage waveform and record them in Table 12-2. Measure the DC output voltage with the DMM and record it in Table 12-2.

Compare Input and Output Voltages (with capacitor filter)

BOTH   CHOP or ALTernate (*best*)   triggered INTernal on CH1 signal

CH 1: e_{supply} GND line centered   DC Coupling

CH 2: v_{out} GND line centered   DC Coupling   Not INVerted

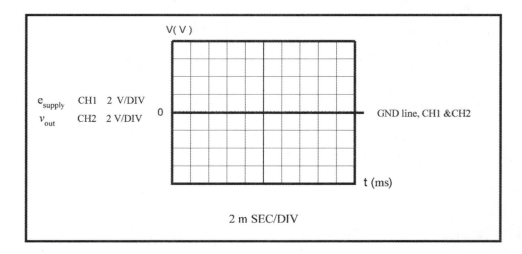

FIGURE 12-3   Half-Wave Rectifier *with* Filter Capacitor

6. Select CH2 and place CH2 in the AC mode to more accurately display the AC voltage riding on the DC offset created. *Increase the sensitivity to observe the largest possible waveform.* Sketch this AC voltage in Figure 12-4. Measure the peak-to-peak voltage of the AC part of the output voltage (called the *ripple voltage*). Record the peak-to-peak ripple voltage in Table 12-3. Also, measure this voltage with the DMM in the AC mode (this measures the RMS of only the AC voltage).

Output AC Ripple Voltage. See Figure 12-4.

CH2 only   triggered INT on CH1 signal (leave CH1 connected)

CH 2: v_{out} GND line centered   AC Coupling   Not INV

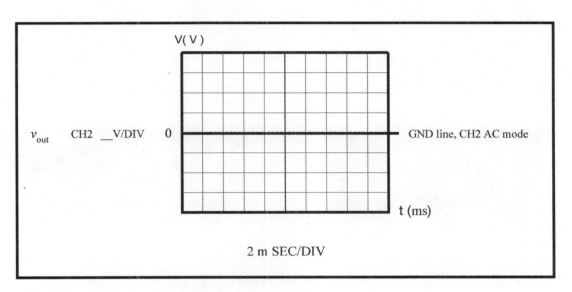

FIGURE 12-4 AC Part of Output Voltage (called *ripple voltage*)

Table 12-3. Ripple Voltage Measurements for Figure 12-1 (AC only values)

Parameter		Expected Value	Measured Value	Percent Difference
$v_{out\ rms}$	DMM RMS	30 mV	🛑	
$v_{out\ pp}$	SCOPE	100 mV	🛑	

🛑 Instructor sign-off of Tables 12-2, 12-3, and oscilloscope displays_____

Observations ▪

1. Compare the output-voltage wave shape to that of the supply voltage as observed in Figure 12-2.

_____

_____

2. Compare the output-voltage wave shapes with and without the capacitor.

_____

_____

3. Compare the DC voltages of e_{in} and v_{out} with the capacitor, and v_{out} without the capacitor.

_____

_____

4. If you were using this circuit to build a DC power supply, which circuit would have the flattest (closest to pure DC) output?

☐ Without capacitor ☐ With capacitor

5. Which basic wave shape best describes the ripple voltage (Figure 12-4)?

☐ Sinusoid ☐ Triangle ☐ Square

6. Calculate the RMS ripple voltage based upon the basic wave shape formula.

7. Is the calculated RMS ripple voltage near to the measured value? ☐ Yes ☐ No

Procedure 12-2

BJT Amplifier

SAMPLE CALCULATIONS

Review Exercise 8 in which this voltage-divider bias circuit was analyzed for its DC voltages. Retrieve the "Expected" DC voltages from that experiment and record them in Table 12-4.

Calculate the expected AC RMS voltages of the base voltage v_b, the collector voltage v_c, and the emitter voltage v_e and record them in Table 12-5 (see the table for formulas). Calculate the expected AC peak-to-peak voltages of the base voltage v_b, the collector voltage v_c, and the emitter voltage v_e and record them in Table 12-6. (see the table for formulas).

Table 12-4 DC Bias Voltages from Previous Experiment

DC Parameter (DMM DC)	Expected DC Values	Measured DC Values
V_{CC}	20 V	
V_B		
V_E		
V_C		

Table 12-5 AC Signal RMS Voltages

AC Parameter (DMM AC)	Expected RMS Values	Measured RMS Values	
e_{in}	141 mV		STOP
$V_b \ (\approx e_{in})$			STOP
$V_e \ (\approx v_b)$			STOP
$V_c \ (\approx 1.75 \times v_b)$			STOP

STOP Instructor sign-off of measured values of Table 12-5 _____

Table 12-6 AC Signal Peak-to-Peak Voltages

AC Parameter (Oscilloscope)	Peak-to-Peak Expected Value	Peak-to-Peak Measured Value	Inverted?
e_{in}			☐ Yes ☐ No
$v_b \ (\approx e_{in})$			☐ Yes ☐ No
$v_e \ (\approx v_b)$			☐ Yes ☐ No
$v_c \ (\approx 1.75 \times v_b)$			☐ Yes ☐ No

Table 12-7 Resistor Data Table for Figure 12-5

Resistor	Nominal Value	Measured Value	Percent Difference	Meet Specs?
R_1	3.3 kΩ			☐ Yes ☐ No
R_2	1 kΩ			☐ Yes ☐ No
R_C	820 Ω			☐ Yes ☐ No
R_E	470 Ω			☐ Yes ☐ No
C	100 μF			☐ Yes ☐ No

1. Measure the values of R_1, R_2, R_C, R_E, and C and record them in Table 12-7.

2. Set up the function generator using CH1 of the oscilloscope to accurately set up the generator frequency, and the DMM to accurately set up the RMS value of the signal. Push in the 0–2 V_{pp} button on the function generator for more precise control of the amplitude. Record actual RMS voltage in Table 12-4. Also, measure its DC voltage and verify that it is close to 0 V_{dc}. (It should be very close to 0 V since the DC OFFSET of the generator should be off.) Any small amount of DC voltage from the generator is filtered out (blocked) by the coupling capacitor.

3. Connect the circuit of Figure 12-5. The circuit layout is shown in Figure 12-6. Measure and record all the DC voltages of Table 12-4.

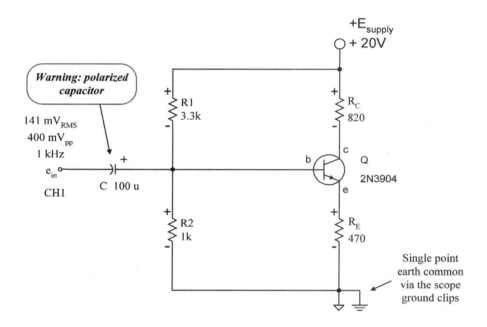

FIGURE 12-5 BJT Amplifier (*pre-lab*)

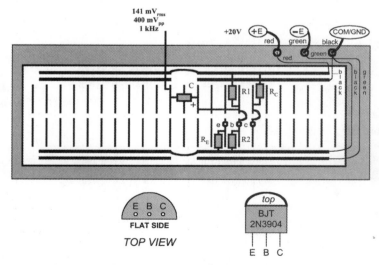

FIGURE 12-6 Suggested Layout for BJT Circuit of Figure 12-5 (*pre-lab*)

Note: BJT flat side facing toward bottom of board. Connect the BJT as indicated with terminals e, b, and c (emitter, base, and collector, respectively).

4. Connect CH1 of the scope to the input signal e_{in} and leave it there throughout this procedure. **Remember: The scope ground leads must be connected to common—a single point in the circuit.** Use the AC mode of the scope to measure and sketch e_{in} on Figure 12-7. Use the *most sensitive appropriate* V/DIV to achieve an accurate oscilloscope reading. Record the peak-to-peak voltage in Table 12-5.

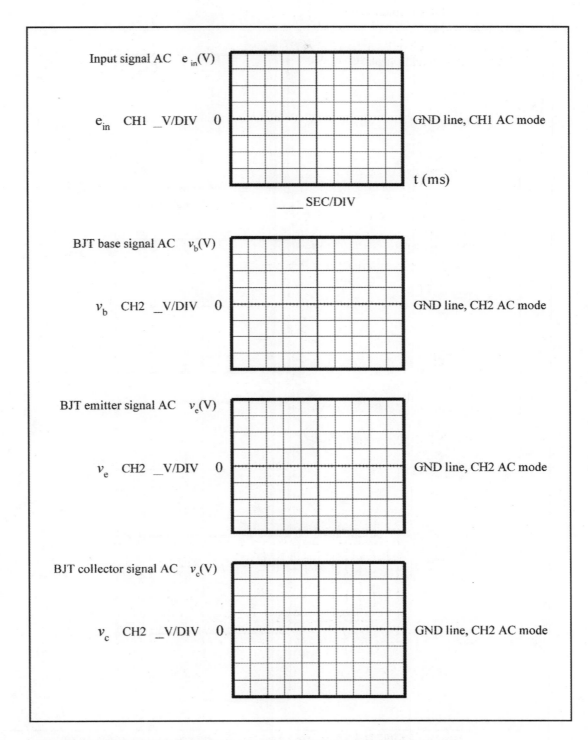

FIGURE 12-7 AC Signal Comparison of e_{in}, v_b, v_e, and v_c with Respect to the Input Signal e_{in}

5. Use CH2 to measure the AC signals v_b, v_c, and v_e. Sketch their waveforms on Figure 12-7 and record their peak-to-peak voltages in Table 12-6. Also, note whether the signal is inverted or not.

6. Leave CH1 connected to e_{in} in the AC mode throughout this procedure. Switch CH2 to the DC coupling mode to measure and sketch the AC signal with DC offset for v_b, v_c, and v_e on Figure 12-8.

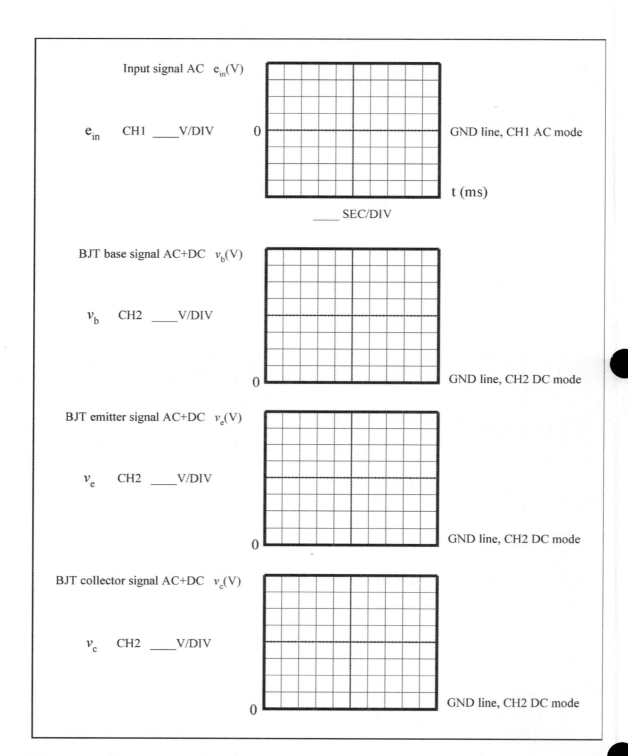

FIGURE 12-8 Total Signal Comparison of e_{in}, v_b, v_e, and v_c with Respect to the Input Signal e_{in}

Observations

1. Calculate the voltage gain from input to emitter.

    Use RMS measured values: $\dfrac{v_e}{e_{in}} =$

    Use peak-to-peak measured values: $\dfrac{v_e}{e_{in}} =$

2. Calculate the voltage gain from input to collector.

    Use RMS measured values: $\dfrac{v_c}{e_{in}} =$

    Use peak-to-peak measured values: $\dfrac{v_c}{e_{in}} =$

3. Was the emitter voltage inverted from the input voltage? (When e_{in} increased did v_e decrease?)
    ☐ Yes   ☐ No

4. Was the collector voltage inverted from the input voltage?   ☐ Yes   ☐ No

Procedure 12-3

Op-Amp Buffer Amplifier

1. Measure the values of R_1 and R_{load} and record them in Table 12-8.

Table 12-8   Resistor Data Table for Figure 12-9

Resistor	Nominal Value	Measured Value	Percent Difference	Meet Specs?
R_1	100 kΩ			☐ Yes ☐ No
R_{load}	1 kΩ			☐ Yes ☐ No

2. Set up the function generator using the oscilloscope to accurately set up the generator frequency and the DMM to accurately set up the RMS value of the signal.
3. Connect the circuit of Figure 12-9(a). Measure and record all of the AC voltages of Table 12-9.
4. Connect the circuit of Figure 12-9(b). Measure and record all of the AC voltages of Table 12-10.

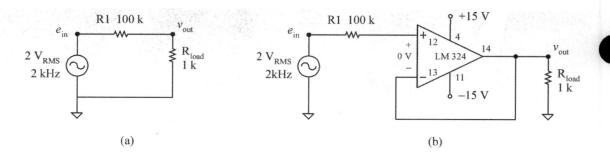

FIGURE 12-9 (**a**) Lossy Circuit and (**b**) Lossy Circuit with Buffer Amplifier (*pre-lab*)

Table 12-9 Lossy Circuit with Significant Signal Attenuation

Parameter	Expected Value	Measured Value
e_{in} DMM	2 V$_{rms}$	
v_{R1} DMM		
v_{out} DMM		

Table 12-10 Lossy Circuit with Buffer Amplifier

Parameter	Expected Value	Measured Value
e_{in} DMM	2 V$_{rms}$	
v_{R1} DMM		
v_{out} DMM		

Observation ■

1. Did the buffer amplifier prevent the signal loss? ☐ Yes ☐ No

Synthesis
Noninverting Op-Amp Amplifier Design

SAMPLE CALCULATIONS

Calculate the resistance value of R_f in the circuit of Figure 12-10 that is required to make the output voltage twice that of the input voltage.

Using previously measured values of headroom for the LM324 op amp, determine the maximum signal output voltage.

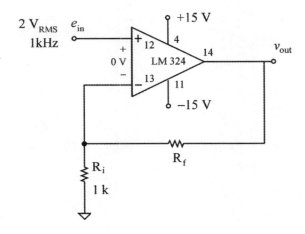

FIGURE 12-10 Amplifier Design Circuit

1. Measure the values of R_i and R_f and record in Table 12-11.

Table 12-11 Resistor Data Table for Figure 12-10

Resistor	Nominal Value	Measured Value	Percent Difference	Meet Specs?
R_i	1 kΩ			☐ Yes ☐ No
R_f				☐ Yes ☐ No

2. Set up the function generator using the oscilloscope to accurately set up the generator frequency and the DMM to accurately set up the RMS value of the signal.

3. Connect the circuit of Figure 12-10. Measure and record all of the AC voltages of Table 12-12.

Table 12-12 Amplifier Circuit Data

Parameter	Expected Value	Measured Value
e_{in} DMM	2 V$_{rms}$	STOP
v_{out} DMM		STOP
maximum $v_{out\ pp}$		STOP

STOP Instructor sign-off of measured values of Table 12-12 _____

4. Increase input signal e_{in} until distortion is observed on the output. Using the oscilloscope, determine the maximum output signal $v_{out\ pp}$ that can be achieved by this amplifier. Record the values in Table 12-12.

Observations ■

1. Was the voltage gain (v_{out}/e_{in}) of your circuit two (2)?   ☐ Yes   ☐ No
2. What was the input voltage when the output voltage was at the maximum undistorted value?

_____

RC Transient Circuits and Transformers

Name: _____   Date: _____

Lab Section: _____ _____   Lab Instructor: _____
                    day   time

Text Reference

DC/AC Circuits and Electronics: Principles and Applications
Chapter 13: Capacitance and Reactance

Materials Required

Function generator 1 Hz–100 kHz (sine, square waveforms; DC offset)
Oscilloscope
1   1N4001 diode
1   Transformer (rated 12.6 V$_{rms}$ @ 1 A)
1   1-kΩ resistor
1   0.1 µF, 47 µF and an unknown value of capacitor

Introduction

In this exercise, you will:

- Examine the sinusoidal and square wave responses of an *RC* circuit.
- Examine a transformer under open and light loading conditions.

The transformer will be used to reduce the 120-Vac line voltage to a level around 15 Vac. The transformer also isolates the common on the low voltage side from the common in the 120-Vac line. A diode will be added to the transformer circuit creating a half-wave rectified AC waveform.

You will then:

- Design a filtering circuit to create a DC level from the half-wave rectified signal.

Pre-Lab Activity Checklist ▪

☐ Perform pre-lab calculations for:

    ☐ Procedure 13-1 to find expected values in Table 13-2.

    ☐ Finding the expected period in Table 13-5.

    ☐ Synthesis procedure to find required capacitor value for 15% voltage ripple.

☐ Read the exercise and have all components prepared for use in the laboratory.

Performance Checklist ▪

☐ Pre-lab completed? 🛑 Instructor sign-off _____

☐ Figure 13-3: τ measurement using oscilloscope.

☐ Figure 13-10 and Table 13-8: Oscilloscope display of filter half-wave rectified signal.

Procedure 13-1.

RC Sinusoidal Waveform Response

SAMPLE CALCULATIONS

Show a sample calculation for the capacitive reactance of a 0.1-µF capacitor at 1 kHz:

X_c (1 kHz) =

---

1. Measure R and C of Figure 13-1 and record them in Table 13-1.

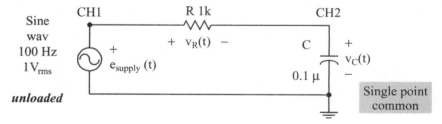

Sine wav 100 Hz 1 V$_{rms}$

e_{supply} (t)

unloaded

CH1   R 1k   CH2

+ v_R(t) −

C

+ v_C(t) −

0.1 µ

Single point common

FIGURE 13-1   *RC* Sinusoid Circuit

TABLE 13-1   Component Data Table for Figure 13-1

Resistor	Nominal Value	Measured Value	Percent Difference
R	1 kΩ		
C	0.1 µF		

2. Connect the circuit of Figure 13-1. After setting up the supply signal e_{supply} using CH1, use CH2 to observe the capacitor voltage (with **INT, CH2** triggering). Observe the capacitor voltage on CH2 to measure peak-to-peak voltage and the DMM to measure the RMS voltage. Record the indicated measurements for the frequencies in Table 13-2.

TABLE 13-2   Data for Figure 13-1

Approximate Generator Frequency (Hz)	Capacitive Reactance (X_c) at this Frequency	v_C DMM Measured RMS Value	v_C Scope Measured PP Value
100 Hz			
1 k			
10 k			
100 k			

Observations

1. With a sinusoidal input, what is the basic shape of the output waveform?

_____

2. As the frequency increased, what happened to the capacitor voltage?

_____

3. Does this affirm the capacitive reactance formula $X_C = \dfrac{1}{2\pi f C}$?   ☐ Yes   ☐ No

4. Does the capacitor act as an open or a short to DC?   ☐ Open ☐ Short
5. Does the capacitor act as an open or a short to high frequencies?   ☐ Open ☐ Short

Procedure 13-2

RC Square Wave Response

1. Measure the resistance of R and capacitance of C and record them in Table 13-3.

TABLE 13-3   Component Data Table for Figure 13-2

Resistor	Nominal Value	Measured Value	Percent Difference
R	1 kΩ		
C	0.1 µF		

2. Calculate the time constant τ from measured values and record it in Table 13-4.

	Nominal Value	Calculated Using Measured R and C	Measured Value Using Oscilloscope	% Difference between Calculated and Measured
TABLE 13-4 Time Constant τ Measurement for Figure 13-2				
$\tau = RC$	100 µS			

3. Connect the circuit of Figure 13-2. Display on **CH2** and sketch the capacitor voltage v_C in Figure 13-3 on page 173.

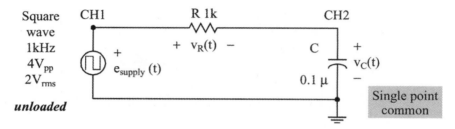

FIGURE 13-2 *RC* Square-Wave Response Circuit

4. Use the scope to measure the time constant of this curve and record it in Table 13-4. Use the 63% technique (up 63% of the change and across τ divisions). Change your SEC/DIV for the most accurate reading of the time constant.

5. Subtract CH1 from CH2 and sketch the resistor voltage v_R in Figure 13-3.

6. Using the resistor voltage curve and Ohm's Law, sketch the current curve.

Observations

1. Which is the smoother curve (no fast changes)?

☐ Capacitor current   ☐ Capacitor voltage

2. How close was the measured time constant τ to the expected value (percent difference)?

_____

3. Is the time constant approximately the same for the capacitor and the resistor voltage curves?

☐ Yes ☐ No

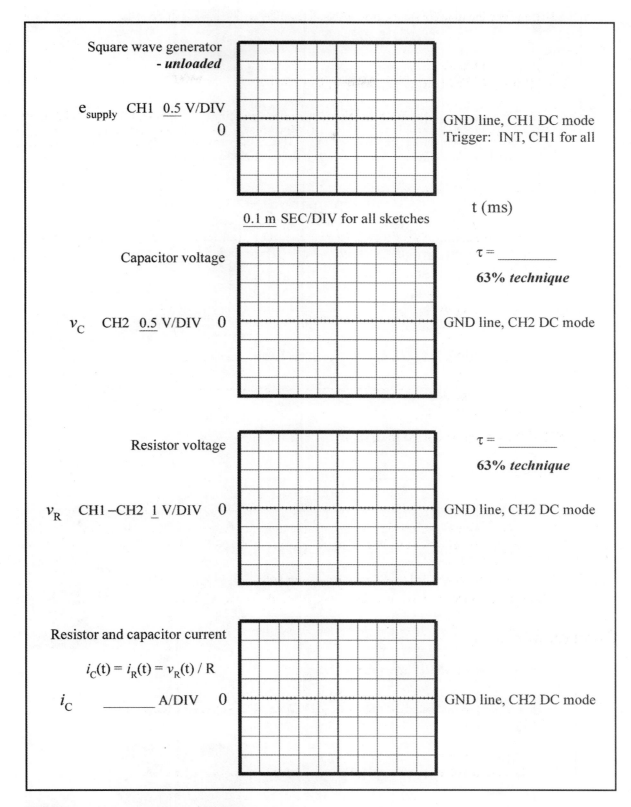

FIGURE 13-3 *RC* Square-Wave Response Waveforms

🛑 Have your instructor verify the τ measurement _____

Procedure 13-3

Transformer

1. Connect the transformer to the AC outlet as shown in Figure 13-4. If the transformer input terminals are exposed, take **extreme caution** to avoid contact with those terminals.

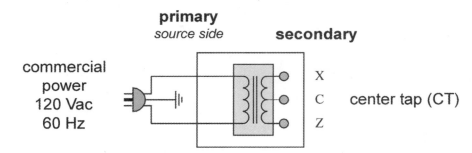

FIGURE 13-4   12.6-V Center-Tapped Transformer (Assume 15 V$_{rms}$ (no load))

2. Using the DMM, measure the transformer secondary DC and RMS voltages and record them in Table 13-5.

TABLE 13-5   Unloaded Transformer: Secondary DC and RMS Voltages

Parameter	Expected Value	Measured DMM Values (RMS)	Measured Scope Values (peak)
V_{XZ}	0 V$_{dc}$		
V_{XC}	0 V$_{dc}$		
V_{ZC}	0 V$_{dc}$		
v_{XZ}	15 V$_{rms}$		
v_{XC}	7.5 V$_{rms}$		
v_{ZC}	7.5 V$_{rms}$		
f frequency	60 Hz		
T period			

3. Connect CH1 of the scope as shown in Figure 13-5 and set to trigger on the **LINE** voltage. If the signal is inverted, be sure the **SLOPE** button is out. If the signal is still inverted, switch your scope leads on the transformer secondary to obtain a noninverted, positive-going waveform as shown.

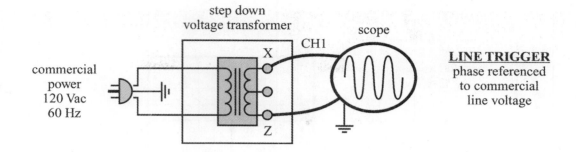

FIGURE 13-5 Measuring the Transformer Secondary Voltage with a Scope

4. See Figure 13-6. Now, Sketch the v_{XZ} waveform in Figure 13-7 on page 176. Measure its peak voltage and its period and record the values in Table 13-5. Calculate the measured frequency from the period and record it in Table 13-5.

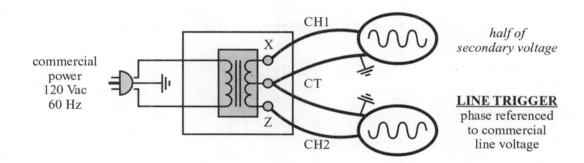

FIGURE 13-6 Measuring the Transformer Center-Tap Voltages

▸ **Note:** Both oscilloscope commons must be connected to the center tap. Connecting the commons to different points will cause the transformer to overheat.

5. Connect the scope common lead to the transformer center tap (CT) and CH1 to the X transformer terminal as noted in Figure 13-6. The X terminal produces the non-inverted voltage waveform. Measure the peak voltage of v_{XC} and record it in Table 13-5. Sketch its voltage waveform in Figure 13-7. Repeat this procedure for v_{ZC}, the inverted, negative-going waveform.

▸ **Note:** If the v_{XZ} display is inverted (initially negative-going), check that the **SLOPE** button is out or in the + position. If the display is still inverted, swap the labeling of your X and Z terminals and reconnect your scope leads to obtain a noninverted, positive-going waveform for v_{XZ}. Also, use your horizontal position control ↔ and your **LEVEL** control to anchor the start of your signal at 0 V. If you have any questions, ask your instructor before continuing.

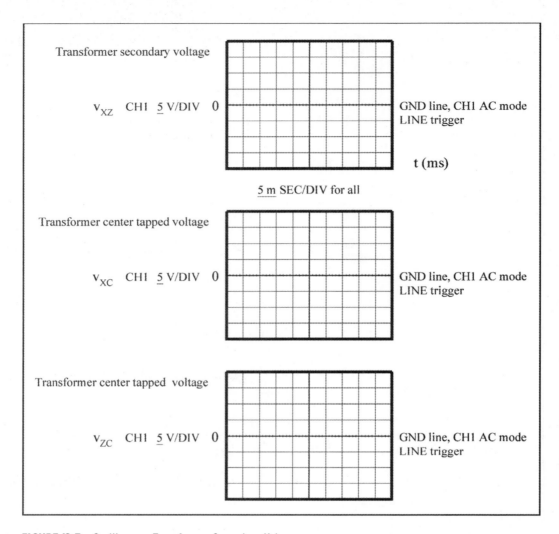

FIGURE 13-7 Oscilloscope Transformer Secondary Voltages

Observations ▪

1. Does the transformer secondary produce DC voltage?   ☐ Yes   ☐ No

2. Does this transformer secondary produce a clean-looking sine wave?   ☐ Yes   ☐ No

3. Based upon your peak and RMS voltage measurements, does the center tap split the secondary voltage in half?   ☐ Yes   ☐ No

4. As v_{XC} goes positive, what does v_{ZC} do?

            ☐ Goes positive (in phase with the line voltage)

            ☐ Goes negative (out of phase with the line voltage)

Procedure 13-4

Transformer with Load

1. Connect the circuit of Figure 13-8. Use **INT, CH1** triggering to view the secondary voltage. Keep CH1 on the secondary of the transformer as the triggering signal throughout the remainder of this exercise.

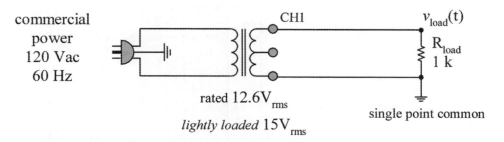

commercial power 120 Vac 60 Hz

CH1

$v_{load}(t)$

R_{load} 1 k

rated 12.6V$_{rms}$

lightly loaded 15V$_{rms}$

single point common

FIGURE 13-8 Transformer Circuit with Resistive Load

2. Measure the DC, RMS, and peak load voltages and record them in Table 13-6. Then sketch the V_{load} oscilloscope waveform in Figure 13-10 on page 178.

TABLE 13-6 Data for Figure 13-8: Transformer Circuit with Resistive Load

Parameter	Expected Value	Measured Value
V_{load}	0 V$_{dc}$	
V_{load}	15 V$_{rms}$	
V_{load}	20 V$_p$	

3. Add the 1N4001 diode as shown in Figure 13-9. Connect CH2 across the load using the same common point as CH1. Make sure that the **INV** and **SLOPE** buttons are disabled. Continue to trigger on **INT, CH1**. Measure the DC, RMS, and peak load voltages and record them in Table 13-7. Sketch the V_{load} waveform in Figure 13-10.

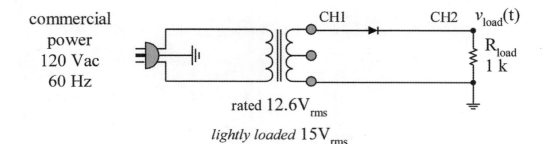

commercial power 120 Vac 60 Hz

CH1 CH2 $v_{load}(t)$

R_{load} 1 k

rated 12.6V$_{rms}$

lightly loaded 15V$_{rms}$

FIGURE 13-9 Rectifier Circuit Using Full Secondary

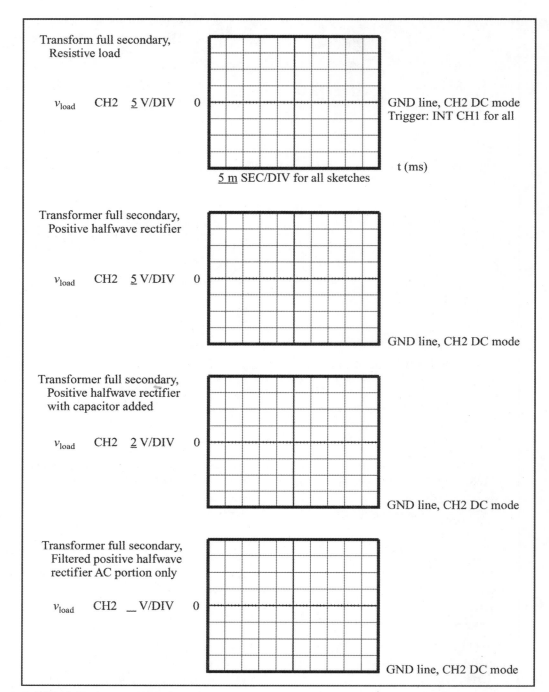

Transform full secondary,
Resistive load

v_{load} CH2   <u>5</u> V/DIV   0

GND line, CH2 DC mode
Trigger: INT CH1 for all

t (ms)

<u>5 m</u> SEC/DIV for all sketches

Transformer full secondary,
Positive halfwave rectifier

v_{load} CH2   <u>5</u> V/DIV   0

GND line, CH2 DC mode

Transformer full secondary,
Positive halfwave rectifier
with capacitor added

v_{load} CH2   <u>2</u> V/DIV   0

GND line, CH2 DC mode

Transformer full secondary,
Filtered positive halfwave
rectifier AC portion only

v_{load} CH2   __ V/DIV   0

GND line, CH2 DC mode

FIGURE 13-10 Sketch of Procedure 13-4 and Synthesis Waveforms

TABLE 13-7 Data Table for Figure 13-9: Rectifier Circuit Using Full Secondary

Parameter	Expected Value	Measured Value
V_{load}	6.3 V_{dc}	
V_{load}	7.7 V_{rms}	
V_{load}	19.3 V_p	

4. Place a 0.1-μF capacitor across the load resistor and observe; but do not sketch the V_{load} waveform.

5. Replace the 0.1-μF capacitor with a 47-μF capacitor and observe; but do not sketch the V_{load} waveform.

WARNING! Be sure electrolytic capacitors are connected with *proper polarity*. Incorrect connection could result in the capacitor exploding.

6. Keep this circuit for use in the Synthesis procedure.

Synthesis
DC Supply Design

SAMPLE CALCULATIONS

To make a half-wave rectified signal fluctuate less, a capacitor is placed across the load. Between peaks of the rectified signal the capacitor voltage decreases following the discharge equation:

$V_C(t) = V_p\, e^{-(t/\tau)}$ where $\tau = R_{\text{load}} \bullet C$

The desired load is a 1-kΩ resistor. We want to select a time constant (τ) that will keep the output voltage from dropping no more than 15% of its peak value. Since the load resistance is fixed at 1 kΩ, we must choose the proper capacitor value.

The period between peaks of the half-wave rectified AC is 16.7 mS (1/60 Hz)

We need to solve for τ where V_C at 16.7 mS is 15% less than V_p (0.85 V_p) or:

$V_C(16.7\text{ mS}) = V_p\, e^{-(16.7\text{ mS}/\tau)} = 0.85\text{ V}_p$

Dividing by V_p yields: $e^{-(16.7\text{ mS}/\tau)} = 0.85$

Solve this equation for the time constant:

$\tau =$

Select a value of capacitor that will yield a time constant close to the τ that you found above. If necessary, combine capacitors in series or parallel to get the proper value.

$C =$

Remove the AC power and add the capacitor of your pre-lab design in parallel with R_{load} of Figure 13-9.

> ⚠️ **WARNING!** Be sure electrolytic capacitors are connected with *proper polarity*. Incorrect connection could result in the capacitor exploding.

Reapply power, then measure and record the voltages in Table 13-8. Sketch the waveforms in Figure 13-10. To view the ripple of the filtered signal, switch CH2 to the AC mode and increase the voltage sensitivity. Measure the peak-to-peak voltage and record it in Table 13-8.

TABLE 13-8 Data for Figure 13-9 with Capacitor Added Across the Load

Parameter	Expected Value	Measured Value
$V_{\text{load dc}}$	8.5 V_{dc}	
$V_{\text{load ac}}$	330 mV_{rms}	
$V_{\text{load max}}$	9.3 V_p	🛑
$V_{\text{load pp}}$	1.1 V_{pp}	🛑

🛑 Have your instructor verify voltage of Figure 13-10 and Table 13-8 _____

Observations ▪

1. What was the effect of adding the 0.1-µF capacitor to the loaded half-wave rectified circuit?

_____

_____

_____

_____

2. Compare the DC voltages in Tables 13-6, 13-7, and 13-8.

_____

_____

_____

_____

3. Compare the RMS in Tables 13-6, 13-7, and 13-8.

_____

_____

_____

_____

4. Compare the peak voltage in Tables 13-6, 13-7, and 13-8.

_____

_____

_____

_____

5. Does the filter capacitor significantly improve the DC load voltage?   □ Yes   □ No
6. Does the filter capacitor decrease the AC RMS voltage?   □ Yes   □ No
7. Calculate the capacitor value required to provide a 5% ripple with a 1-kΩ load.

Characteristic Curves of Rectifier and Zener Diodes

Name: _____   Date: _____

Lab Section: _____ _____   Lab Instructor: _____
                 day   time

Text Reference

DC/AC Circuits and Electronics: Principles and Applications
Chapter 3: Resistive Circuits
Chapter 12: Waveforms

Materials Required

Triple power supply (2 @ 0–20 volts DC; 1 @ 5 volts DC)
Oscilloscope
Function generator
Three-prong-to-two-prong AC plug adapter
1   2.2-kΩ resistor
1   1N753 zener diode
1   1N4001 rectifier or equivalent

Introduction

The rectifier diode is a silicon-based semiconductor that will allow current to flow through it when it has a positive voltage of at least 0.7 V applied to its anode relative to its cathode. Essentially, the rectifier diode will not conduct if it has a voltage that is positive on its cathode relative to its anode (reverse bias).

The zener diode is a device that will conduct when it is reverse biased provided that the voltage applied is equal to or greater than the zener voltage rating. Zener diodes are almost exclusively used in the reverse-bias mode.

In this exercise, you will:

* Locate and read a manufacturer's part specification. (These specifications can be found at sites such as: **fairchildsemiconductor.com, gensemi.com, motorola.com,** and **nationalsemiconductor.com**.)
* Plot voltage-versus-current characteristics of a rectifier diode
* Plot voltage-versus-current characteristics of a zener diode
* Use the oscilloscope in the X-Y mode.

Pre-Lab Activity Checklist ■

☐ Find all specifications in Tables 14-1 and 14-2.
☐ Calculate the expected values in Tables 14-4, 14-5, 14-6, 14-7, and 14-8.

Performance Checklist ■

☐ Pre-lab completed?   🛑 Instructor sign-off _____
☐ Table 14-4 Demonstrate −10-V measurement
☐ Table 14-7 Demonstrate 2-V measurement
☐ Table 14-8 Demonstrate − 9-V measurement

Procedure 14-1

Device Specifications

1. Read the part number of your rectifier diode. Fill in each of the blanks in Table 14-1. Utilize the textbook appendix or data books (available online) to find the data specifications for your rectifier diode.

TABLE 14-1 Rectifier Diode Specifications

Diode Specification	Diode Rating for Your Rectifier Diode
V_{RRM} V_{RWM} V_R	
V_{RSM}	
$V_{R(rms)}$	
I_0	
I_{FSM}	
T_J, T_{stg}	
V_F	
$V_{F(AV)}$	
I_R	
$I_{R(AV)}$	

2. Record the part number of your zener diode in Table 14-2. Fill in each of the blanks in Table 14-2. Utilize the textbook appendix or data books (available online) to find the data specifications for your zener diode.

3. Calculate the expected $V_{Z(max)}$ and $V_{Z(min)}$ values. Assume a ±10% tolerance, where $V_{Z(max)} = V_{Z(typ)} + 0.1V_{Z(typ)}$ and $V_{Z(min)} = V_{Z(typ)} - 0.1V_{Z(typ)}$. Record the minimum and maximum values for V_Z in Table 14-2.

TABLE 14-2 Zener Diode Specifications

Diode Specification	Zener Diode Part Number-_____ Diode Rating		
P_D			
V_Z	(min)	(typ)	(max)
I_{ZT}			
Z_{ZT}			
I_{ZM}			
I_R			

SAMPLE CALCULATIONS

Show your calculations for:

$V_{z(min)} =$

$V_{z(max)} =$

Procedure 14-2

Characteristics of a Short, a Resistor, and an Open

1. Measure all resistances of R_S and R_X and record them in Table 14-3 on page 186. **Remember: You must accurately use your ohmmeter to measure very low resistance values by proper use of the relative button or by taking into account the shorted probe contact resistance.**

   Always measure any resistors used in a circuit to ensure that they are within tolerance.

   ❏ **Note:** A percentage error calculation is not mathematically possible for either the short or the open. Represent the infinite (open) resistance with the very large value (1E12 Ω) if using a spreadsheet since there is no means to enter infinity.

TABLE 14-3 Resistor Values

Resistor	Rated R_X	Measured R_X	Error %
R_S	1 kΩ		
R_X	0 Ω short		
R_X	2.2 kΩ		
R_X	∞ Ω open		

2. Build the circuit of Figure 14-1 with R_X a short (piece of wire). E_S is a *positive* power supply relative to COM.

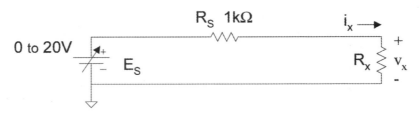

FIGURE 14-1 Series Circuit with R_x Being a Short

3. For each *positive* E_S entry in Table 14-4, adjust the voltage source to approximately that E_S value. Measure and record the device voltage V_X and device current I_X. Start with 0 V and proceed to the most positive E_S.

TABLE 14-4 Operating Points to Generate a Characteristic Curve of a Short

E_S	Expected V_X	Expected I_X	Measured V_X	Measured I_X
20 V				
10 V				
0 V				
−10 V			STOP	STOP
−20 V				

➔ **Note:** Use the power supply panel meter to set E_S. The accuracy of E_S is not important; however, accuracy of the corresponding voltage-current operating point values is important.

STOP Instructor sign-off of measured values in Table 14-4 _____

SAMPLE CALCULATIONS

Show your calculations for:

$V_x =$

$I_x =$

4. Construct the circuit of Figure 14-2. Note that E_S is now a *negative* power supply relative to COM.

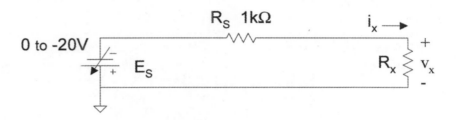

FIGURE 14-2 Reverse-Biased Device Using Negative Supply

5. For each *negative* E_S entry in Table 14-4, adjust the voltage source to approximately that E_S value. Measure and record the device voltage V_X and device current I_X. Start with 0 V and proceed to the most *negative* $-E_S$. Again, use the power supply panel meter to *approximately* set up the voltage source, then measure the V_X and I_X pairs accurately.

6. Replace R_X with a 2.2-kΩ resistor and complete Table 14-5.

TABLE 14-5 Operating Points to Generate a Characteristic Curve of a 2.2-kΩ Resistor

E_S	Expected V_X	Expected I_X	Measured V_X	Measured I_X
20 V				
10 V				
0 V				
−10 V				
−20 V				

SAMPLE CALCULATIONS

Show your calculations for:

$V_x =$

$I_x =$

7. Replace R_x with an open and complete Table 14-6.

TABLE 14-6 Operating Points to Generate a Characteristic Curve of an Open

E_s	Expected V_x	Expected I_x	Measured V_x	Measured I_x
20 V				
0 V				
0 V				
−10 V				
−20 V				

SAMPLE CALCULATIONS

Show your calculations for:

$V_x =$

$I_x =$

Observations

1. When R_x was a short, what voltage V_x did you measure? ________
   Why was it that value? (*Hint:* Use Ohm's Law to quantify your answer.)

   ____

   ____

2. When R_x was a short, did the measured voltage V_x change as the current I_x changed?
   ☐ Yes ☐ No Explain your answer.

   ____

   ____

3. When R_x was a 2.2-kΩ resistor, what value of V_x and I_x did you measure at $E_s = -10$ V?

   ____

   ____

4. When R_x was a 2.2-kΩ resistor, did the measured voltage V_x change as the current I_x changed?
   ☐ Yes ☐ No Explain why or why not.

   ____

   ____

5. When R_x was an open, what value of V_x and I_x did you measure when E_s was +20 V? ________ Why? Use Kirchoff's Voltage Law to quantify your answer.

   ____

   ____

Procedure 14-3

Characteristic Curve of a Rectifier Diode

1. Use the diode checker on your DMM to check the diode's forward conduction. The forward-biased diode checker has the symbol ⇥. Refer to the DMM instrument manual to determine how your diode checker works.

   Forward-biased measurement possibilities include:
   - Beeps if the diode passes a forward-bias test (red lead on anode, black on cathode).
   - Displays the diode forward-biased voltage drop.
   - Displays the diode's DC forward resistance on the ohmmeter.

   The ohmmeter can be used to check the reverse-biased diode (which should indicate an open—really a very large reverse resistance that is not measurable with the DMM).

2. Build the test circuit of Figure 14-3 to measure data points for the forward-biased characteristic curve of the rectifier diode.

FIGURE 14-3 Forward-Biased Rectifier Diode

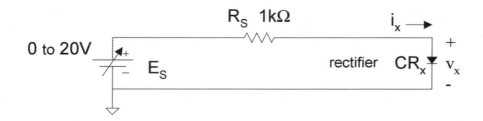

3. For each *positive* E_S entry in Table 14-7, adjust the source voltage approximately to that positive E_S value and measure/record the device voltage V_X and device current I_X. Start with 0 V and proceed to the most positive E_S. The negative values will be measured and recorded in the next step. Measuring voltage and current simultaneously is a great time saver on this one.

TABLE 14-7   Operating Points for Generating a Rectifier Diode Characteristic Curve

E_s	Expected V_X	Expected I_X	Measured V_X	Measured I_X
20.0 V				
10.0 V				
2.0 V			⑤ STOP	⑤ STOP
1.0 V				
0.7 V				
0.5 V				
0.0 V				
−1.0 V				
−2.0 V				
−10.0 V				
−20.0 V				

⑤ STOP Instructor sign-off of measured values in Table 14-7 _____

SAMPLE CALCULATIONS

Show your calculations for:

$V_x =$

$I_x =$

4. Construct the circuit of Figure 14-4. Note that E_S is now a *negative* power supply relative to COM, and the device is reverse biased.

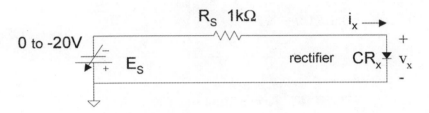

FIGURE 14-4 Reverse-Biased Rectifier Diode

5. For each *negative* $-E_S$ entry in Table 14-7, adjust the source voltage approximately to that negative E_S value and measure/record the device voltage V_X and device current I_X. Start with 0 V and proceed to the most negative voltage.

Observation

1. Compare the magnitude of the measured I_x values when E_s was +20 V to when E_s was −20 V. Are they different or nearly identical? Why?

_____

_____

_____

_____

Procedure 14-4

Characteristic Curve of a Zener Diode

1. Use the diode checker ⇥ on your DMM to check the diode's forward conduction. Check the reverse-bias direction with the DMM ohmmeter; it should indicate an open.

    ⊐ **Note:** The reverse-biased zener will break down and conduct if the ohmmeter applies a test signal greater than V_Z.

2. Construct the test circuit of Figure 14-5 to measure data points for the forward-biased characteristic curve of the zener diode.

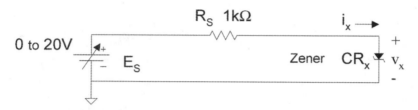

FIGURE 14-5 Forward-Biased Zener Diode

3. For each *positive* E_S entry in Table 14-8, adjust the source voltage to that positive E_S value and measure/record the device voltage V_X and device current I_X. Start with 0 V and proceed to the most positive E_S.

SAMPLE CALCULATIONS

Show your calculations for:

$V_X =$

$I_X =$

---

4. Construct the circuit of Figure 14-6. Note that E_S is now a *negative* power supply relative to COM and the diode is reverse biased.

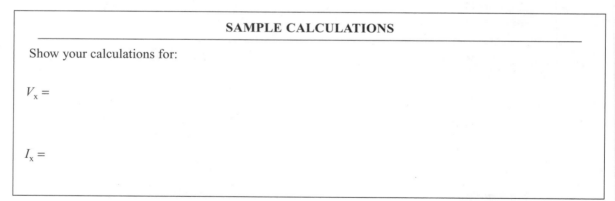

FIGURE 14-6 Reverse-Biased Zener Diode (normal operation)

TABLE 14-8 Operating Points for Generating a Zener Diode Characteristic Curve

E_S	Expected V_X	Expected I_X	Measured V_X	Measured I_X
20.0 V				
10.0 V				
2.0 V				
1.0 V				
0.7 V				
0.5 V				
0.0 V				
−2.0 V				
−5.0 V				
−6.0 V				
−7.0 V				
−9.0 V			STOP	STOP
−15.0 V				
−20.0 V				

STOP Instructor sign-off of measured values in Table 14-8 _____

5. For each *negative* E_S entry in Table 14-8, adjust the source voltage approximately to that negative E_S value and measure/record the device voltage V_X and device current I_X. Start with 0 V and proceed to the most negative value.

Observations ■

1. Compare the value of I_x when E_s was 2.0 V to when the value of E_s was −2.0 V. Are they different or nearly identical? Why?

_____

_____

_____

_____

2. Compare the value of I_x when E_s was 20.0 V to when the value of E_s was -20.0 V. Are they different or nearly identical? Why?

_____

_____

_____

_____

Procedure 14-5

Observing the Zener Diode Characteristic Curve Using the Oscilloscope X-Y Mode

1. Place a three-to-two prong AC adapter in the line of the function generator you will be using in Figure 14-7. This allows the generator chassis ground to be different from the oscilloscope ground.

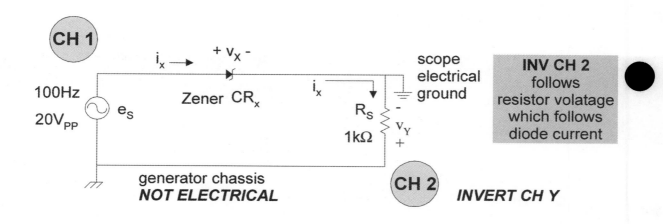

FIGURE 14-7 Circuit to Observe X-Y Trace of the Zener Diode Characteristic Curve

2. Construct the circuit of Figure 14-7. The oscilloscope CH1 (x-axis) traces the voltage v_X across the device and CH2 (y-axis) traces the voltage v_Y across the series resistor (the same shape as the circuit and diode current i_X). Observe the oscilloscope in the X-Y mode with CH2 **INV**erted to produce the correct reference polarity. The oscilloscope will display V_Y versus V_X. By Ohm's Law, you may divide CH2 (y-axis) "V/DIV" by R_S so that the vertical scale translates to the device current scale. Thus, the sketch becomes I_{device} versus V_{device}. The oscilloscope must be used in the DC mode since the AC mode uses an internal RC circuit that filters out DC voltage and affects the observed signal.

3. Observe the characteristic curve and sketch it on Figure 14-8. Adjust the sensitivity of CH2 to scale the y-axis to current.

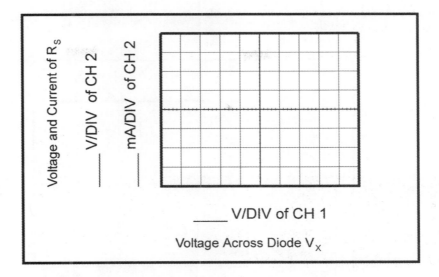

FIGURE 14-8 X-Y Trace of the Zener Diode Characteristic Curve

4. Reverse the CH2 invert switch so that CH2 of the oscilloscope is operating normally. Also, take the oscilloscope out of X-Y mode and put it into alternate mode to display both CH1 and CH2.

5. Remove the three-to-two prong AC adapter.

Synthesis

Using the Diode as a Rectifier

Construct the circuit of Figure 14-9 with a sine wave input. The rectifier diode allows current to flow in only one direction in the circuit. This is called *rectification*.

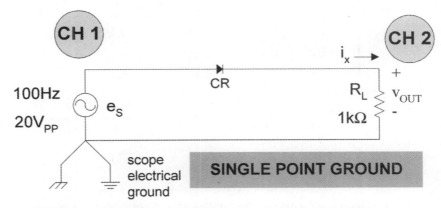

FIGURE 14-9 Circuit to Observe Rectifier Circuit with Positive Output Voltage

Observe the input signal sine wave, e_S, on Channel 1 of the oscilloscope and the output signal v_{out} on Channel 2 of the oscilloscope. Use the dual-trace mode of the oscilloscope to observe both signals simultaneously. Trigger the oscilloscope internally on CH1 to make the phase of the output relative to the input signal. Sketch the waveforms observed in Figure 14-10.

Reverse the direction of the rectifier diode. Observe and sketch the dual-trace oscilloscope waveforms in Figure 14-11.

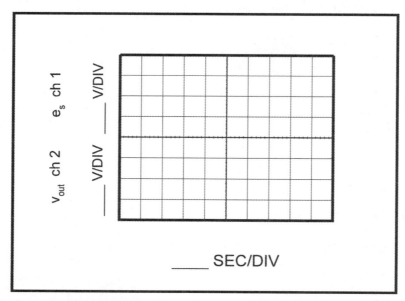

FIGURE 14-10 Rectifier Circuit with Positive Output Voltage: Dual Trace

e_s Top half of display
v_{out} Bottom half of display

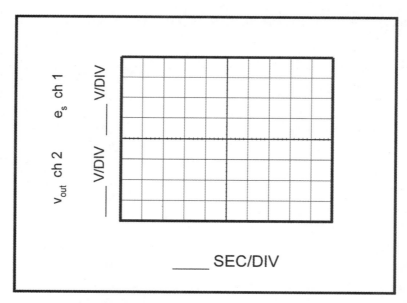

FIGURE 14-11 Rectifier Circuit with Negative Output Voltage: Dual Trace

e_s Top half of display
v_{out} Bottom half of display

Computer Activity

Characteristic Curves of a Short, a Resistor, and an Open

Using Data Tables 14-4, 14-5, and 14-6 and the spreadsheet graphing capabilities, create each characteristic curve on a single graph as follows (these instructions are for Microsoft Excel®):

1. Set up a single spreadsheet table as shown in Table 14-9, using measured data from Tables 14-4, 14-5, and 14-6.

TABLE 14-9   Spreadsheet Format

Measured Voltages V_x	Table 14-4 Measured Currents I_x	Table 14-5 Measured Currents I_x	Table 14-6 Measured Currents I_x
a	a		
a	a		
a	a		
b		b	
b		b	
b		b	
c			c
c			c
c			c

2. Highlight the numerical data (shaded area).
3. Click on the "ChartWizard" icon (border highlighted).
4. Box an area where the chart is to be displayed.
5. Select the **XY (scatter plot)** option (by double-clicking the icon or clicking the icon and "Next").
6. Select the #2 plot option.
7. Select "Next" (X-axis data is in the first column).
8. Chart title: "Characteristic Curves," Category (X): "x - volts." Value (Y): "y - mA"; then click "Finish."
9. The resulting graph should look similar to Figure 14-12.

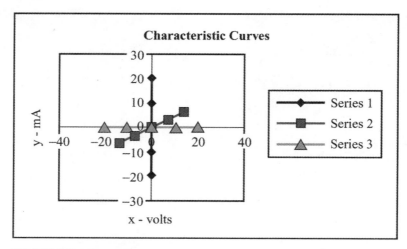

FIGURE 14-12　Sample Graph

10.　Double-click on the chart to highlight it for editing.

11.　Double-click on vertical curve and modify the "Names and Values" tab to 'Name' **short**.

12.　Double click on diagonal curve and modify the "Names and Values" tab to 'Name' **2.2k.**

13.　Double-click on horizontal curve, modify the "Names and Values" tab to 'Name' **open** and modify the "Patterns" tab to color everything **black** so its print is more distinctive.

14.　The chart should now look similar to Figure 14-13.

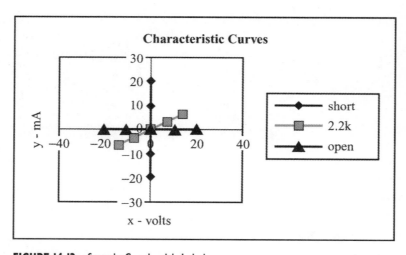

FIGURE 14-13　Sample Graph with Labels

15.　Click on a cell outside of the chart to exit the *edit chart* mode.

16.　Print out and attach your table and chart to your lab report.

Observations

1. The characteristic curve of a short is:

   ☐ Vertical ☐ Diagonal ☐ Horizontal

2. Find the slope of the characteristic curve for a short.

3. The characteristic curve of an open is:

   ☐ Vertical ☐ Diagonal ☐ Horizontal

4. Find the slope of the characteristic curve for an open.

5. The characteristic curve of a resistor is:

   ☐ Vertical ☐ Diagonal ☐ Horizontal

6. Find the slope of the characteristic curve for any resistor. *Hint:* Calculate slope $m = y/x$ and compare with $1/R$.

Characteristic Curve of a Rectifier Diode

Using a spreadsheet and Table 14-6, plot the rectifier characteristic curve. Attach it to your report.

Observation

1. Based on the rectifier curve, estimate the forward knee voltage of your diode (show the point on your characteristic curve).

   $V_{knee} =$ _____

Characteristic Curve of a Zener Diode

Using a spreadsheet and Table 14-7, plot the zener characteristic curve. Attach it to your report.

Observation

1. Based on the zener curve, estimate the zener breakdown voltage. Show the point on your characteristic curve.

   $V_{breakdown} =$ _____

Timer, Differentiator, and Integrator Circuits

Name: _____   Date: _____

Lab Section: _____ _____   Lab Instructor: _____
　　　　　　　　　day　　　　　　　　time

Text Reference

DC/AC Circuits and Electronics: Principles and Applications
Chapter 15: Wave Shaping and Generation

Materials Required

　　Triple power supply (2 @ 0–20 volts DC; 1 @ 5 volts DC)
　　Oscilloscope
　　Function generator
　　2　each 1-kΩ, 100-kΩ resistor
　　1　10-kΩ, 100-kΩ resistor
　　1　NE555 integrated circuit
　　1　LM741 integrated circuit
　　1　each 0.01-μF, 0.1-μF, 0.47-μF, 1.0-μF, and 22-μF (min. 35 WVdc) capacitors

Introduction

In this exercise, you will:

- Experimentally demonstrate the operation of the astable multivibrator using the 555 IC
- Experimentally demonstrate the operation of the op-amp integrator circuit using the uA741 IC
- Experimentally demonstrate the operation of the op-amp differentiator using the uA741 IC
- Design a circuit using the 555 IC as a monostable multivibrator.

Pre-Lab Activity Checklist

☐ Find the expected values in Table 15-2.
☐ Find the expected values in Table 15-3.
☐ Calculate the expected values in Table 15-4.
☐ Build the circuit of Figure 15-1.

Performance Checklist

☐ Pre-lab completed?   **STOP** Instructor sign-off _____

☐ Demonstrate T_W and D measurement for 1.0-μF capacitor in Table 15-2.
☐ Demonstrate the minimum and maximum values for the 1.0-μF capacitor in Table 15-3.
☐ Demonstrate the minimum and maximum values for the 1.0-μF capacitor in Table 15-4.

Procedure 15-1

Component Measurement

1. Measure the values of the resistors and capacitors used in this exercise and record them in Table 15-1.

TABLE 15-1 Measured Component Values

Rated Value	Measured Value
1 kΩ #1	
1 kΩ #2	
10 kΩ	
100 kΩ #1	
100 kΩ #2	
0.01 μF	
0.1 μF	
0.47 μF	
1 μF	
22 μF	

Procedure 15-2

The Astable Multivibrator

1. Connect the astable multivibrator circuit of Figure 15-1. Using the oscilloscope observe the output and sketch the waveform in Figure 15-2.

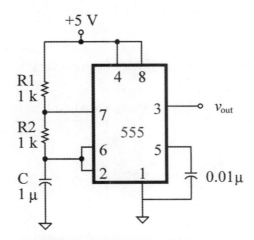

FIGURE 15-1 Astable Multivibrator (*pre-lab*)

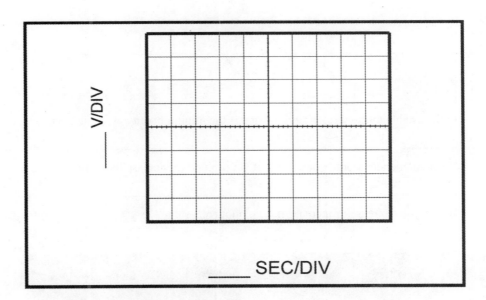

FIGURE 15-2 Astable Multivibrator Circuit Waveforms

2. Measure and record the output period and duty cycle in Table 15-2 on page 204.

3. Substitute the 0.47-μF capacitor for C_1 in Figure 15-1. Measure and record the period and duty cycle in Table 15-2.

4. Substitute the 0.1-μF capacitor for C_1 in Figure 15-1. Measure and record the period and duty cycle in Table 15-2.

TABLE 15-2 Astable Multivibrator Circuit Data

Value of C_1	T_W Expected	D Expected	T_W Measured	D Measured
1.0 μF			(STOP)	(STOP)
0.47 μF				
0.1 μF				

(STOP) Instructor sign-off of measured values in Table 15-2 _____

5. Leave this circuit built. It may be modified for use in the synthesis exercise.

SAMPLE CALCULATIONS

Show your pre-lab calculations for:

$T_W =$

$D =$

Observations ■

1. In the astable multivibrator circuit, what is the minimum duty cycle that is practically possible?

   _____

2. What component would you change if you wanted to increase the duty cycle of the output signal?

   _____

3. Would you decrease or increase this component's value? ☐ Increase ☐ Decrease

Procedure 15-3

The Integrator

1. Construct the op-amp integrator circuit of Figure 15-3.
2. Apply a 10 V_{pp}, 1-kHz square wave to the input of the integrator.

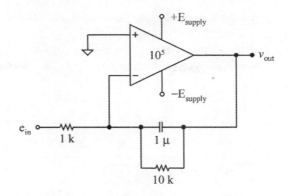

FIGURE 15-3 The Op-Amp Integrator

3. Using the oscilloscope (AC mode, 2 V/DIV, 200 μS/DIV), observe the output waveform. Sketch the waveform in Figure 15-4.

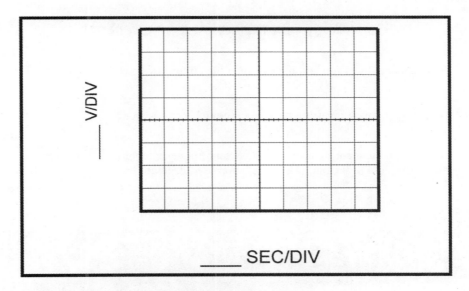

FIGURE 15-4 The Op-Amp Integrator Circuit Waveforms

4. Measure and record the positive peak and negative peak values of the output and record them in Table 15-3.

TABLE 15-3 Integrator Circuit Data

Value of C_1	V_{max} Expected	V_{min} Expected	V_{max} Measured	V_{min} Measured
1.0 μF			(STOP)	(STOP)
0.47 μF				
0.1 μF				

(STOP) Instructor sign-off of measured values in Table 15-3 _____

5. Substitute the 0.47-µF capacitor for C_1 in Figure 15-3. Measure and record the positive peak and negative peak values of the output and record them in Table 15-3.

6. Substitute the 0.1-µF capacitor for C_1 in Figure 15-3. Measure and record the positive peak and negative peak values of the output and record them in Table 15-3.

SAMPLE CALCULATIONS

Show your sample calculation for:

$V_{max} =$

Observations

1. Is the integrator circuit a positive feedback or a negative feedback device?

☐ Positive ☐ Negative

2. How can one, by examination, determine if a circuit uses positive or negative feedback?

_____

3. Is the integrator an inverting device or a noninverting device?

☐ Inverting ☐ Noninverting

4. If you maintain a constant input voltage and frequency applied to the integrator but increase the value of the capacitor, will the output voltage increase or decrease?

☐ Increase ☐ Decrease

Procedure 15-4

The Differentiator

1. Construct the op-amp differentiator circuit of Figure 15-5.

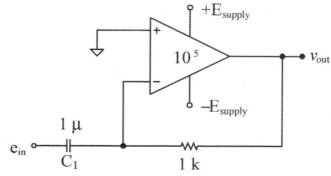

FIGURE 15-5 Op-Amp Differentiator

2. Apply a 4-V$_{pp}$, 1-kHz triangle wave to the input of the differentiator.

3. Observe the output signal with the oscilloscope, then sketch the waveform of the output in Figure 15-6.

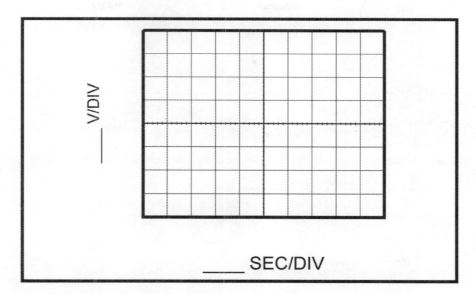

FIGURE 15-6 Op-Amp Differentiator Waveforms

4. Measure and record the maximum and minimum output values and record them in Table 15-4.

TABLE 15-4 Differentiator Circuit Data				
Value of C_1	V_{max} **Expected**	V_{min} **Expected**	V_{max} **Measured**	V_{min} **Measured**
1.0 μF			🛑	🛑
10 μF				
0.1 μF				

🛑 Instructor sign-off of measured values in Table 15-4 _____

5. Substitute the 0.47-μF capacitor for C$_1$ in Figure 15-5. Measure and record maximum and minimum output values and record them in Table 15-4.

6. Substitute the 0.1-μF capacitor for C$_1$ in Figure 15-5. Measure and record the maximum and minimum output values and record them in Table 15-4.

SAMPLE CALCULATIONS

Show your sample calculation for:

$V_{min} =$

Observations ◼

1. Describe the differences in the component placement between the differentiator and integrator.

   _____

   _____

   _____

2. Is the differentiator a positive feedback device or a negative feedback device?

   ☐ Positive   ☐ Negative

Synthesis

The Monostable Multivibrator

1. Choose the R and C values for a monostable multivibrator circuit of Figure 15-7 that will yield a pulse width of approximately 1.2 seconds.

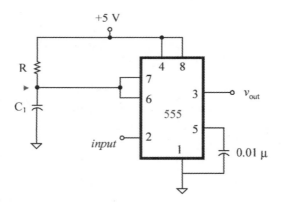

FIGURE 15-7   The 555 Monostable Multivibrator Circuit

2. Build the circuit using the values of R and C that you selected.
3. Connect the input at pin 2 to a switch. Set the switch to the logic 1 or high position.
4. Set the oscilloscope to 0.5-S/DIV sweep rate and 2-V/DIV vertical sensitivity.

5. Momentarily connect the input on pin 2 to a logic low (0) and then return it to the logic high (1) position.

6. Using the oscilloscope, measure and record the output waveform pulse width at pin 3. Repeat the process of steps 5 and 6 until you have an accurate measurement of the output pulse width. Record your output pulse width in Table 15-5.

TABLE 15-5 Monostable Multivibrator Pulse Width

Expected Value of T_W	Measured Value of T_W
1.2 seconds	

Observation ■

1. In the monostable multivibrator circuit using the 555 timer, is the output pulse initiated by a positive-going or a negative-going transition on the input?

☐ Positive ☐ Negative

Computer Activity

1. Simulate the circuit of Figure 15-1 and compare the results of the simulation with the actual results obtained.

How are the two results similar? _____

_____

How are they different? _____

_____

2. Simulate the circuit of Figure 15-3 and compare the results of the simulation with the actual results obtained.

How are the two results similar? _____

_____

How are they different? _____

_____

3. Simulate the circuit of Figure 15-5 and compare the results of the simulation with the actual results obtained.

How are the two results similar? _____

_____

How are they different? _____

_____

Relaxation Oscillator and Waveform Generator

Name: _____   Date: _____

Lab Section: _____ _____   Lab Instructor: _____
               day   time

Text Reference ◼

DC/AC Circuits and Electronics: Principles and Applications
Chapter 15: Wave Shaping and Generation

Materials Required ◼

Triple power supply (2 @ 0–20 volts DC; 1 @ 5 volts DC)
1   each 1-kΩ, 1.8-kΩ, 2.2-kΩ, 3.3-kΩ, 4.7-kΩ, 6.8-kΩ, 8.1-kΩ, 10-kΩ, 18-kΩ, and 100-kΩ resistor
1   0.1-µF capacitor
1   10-kΩ single-turn potentiometer
1   uA741 integrated circuit

Introduction ◼

In this exercise, you will:

- Examine a relaxation oscillator for square wave generation
- Examine an *RC* integrator circuit
- Examine a variable-frequency variable waveform generator.

Procedure 16-1

The Square Wave Generator

1. Connect the circuit of Figure 16-1 and measure the output waveform with an oscilloscope. Draw the output waveform in Figure 16-2, for the first set of values in Table 16-1. Record the output period in Table 16-1.

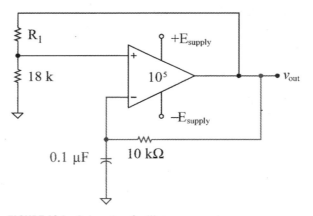

FIGURE 16-1 Relaxation Oscillator

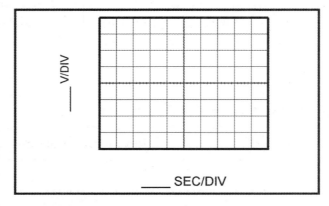

FIGURE 16-2 The Relaxation Oscillator Circuit

TABLE 16-1 Relaxation Oscillator Output

Nominal Value of R_1	Expected Output Period	Measured Output Period
3.3 kΩ		STOP
6.8 kΩ		
8.1 kΩ		
10 kΩ		

STOP Instructor sign-off of measured values in Table 16-1 _____

2. Substitute the second value for R_1 in the circuit and measure and record the output period in Table 16-1.
3. Substitute each of the remaining values for R_1 listed in Table 16-1 and record the resulting period for each.

SAMPLE CALCULATIONS

Show your sample pre-lab calculations for:

T (output period) =

---

4. Place a 3.3-kΩ resistor in series with a single-turn 10-kΩ potentiometer into the circuit for R_1. Set the potentiometer to its maximum value and measure and record the output frequency in Table 16-2.
5. Set the potentiometer to its minimum value and measure and record the output frequency in Table 16-2.
6. Keep this circuit for use in the synthesis procedure.

TABLE 16-2 Oscillator Maximum and Minimum Output Frequency

Potentiometer Setting	Expected Output Frequency	Measured Output Frequency
Maximum		
Minimum		

SAMPLE CALCULATIONS

Show your sample pre-lab calculations for:

$f_{max}=$

$f_{min}=$

Observations

1. Does the value of the trip-point voltage affect the amplitude of the output?　☐ Yes　☐ No
2. What characteristic of the output signal does the trip-point voltage affect?

Synthesis

Variable Frequency Triangle and Square Wave Generator

Calculate the values needed for R_2, R_3, and C_2 of Figure 16-3 that will limit the output of the integrator to one-tenth of the amplitude of its square wave input. Record the values in Table 16-3. Select the capacitor and resistor combination from the parts available for this exercise. Using your relaxation oscillator as a square wave generator, construct the circuit of Figure 16-3 with the value for R_2 and C_2 that you have selected.

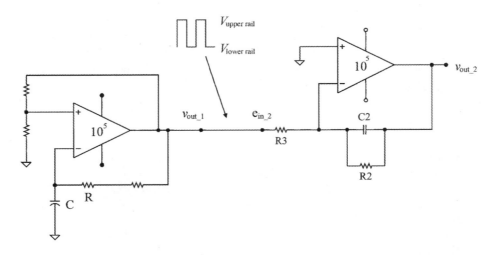

FIGURE 16-3　Variable Frequency Triangle and Square-Wave Generator

TABLE 16-3 Triangle Wave Generator Output

Value of R_2	Value of R_3	Value of C_2	Expected V_{out}	Measured V_{out}

Measure the output of the square wave generator on Channel 1 of the oscilloscope and measure the output of the integrator with Channel 2 of the oscilloscope. Sketch the two waveforms in Figure 16-4 with the 10-kΩ potentiometer adjusted to give the minimum output frequency.

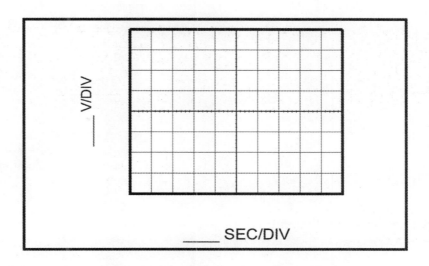

FIGURE 16-4 Square-Wave and Triangle-Wave Outputs

Observations

1. What is the phase relationship between the output of the square wave generator and the output of the triangle wave generator?

2. What is the relationship between the input frequency to the integrator and the output frequency of the integrator?

3. What is the relationship between the input amplitude to the triangle-wave generator and its output amplitude?

4. How could you vary the amplitude of the square-wave generator circuit? What circuit and components would you change?

Computer Activity

1. Simulate the circuit of Figure 16-1. Compare the results of the simulation with the actual results obtained.

   How are the two results similar? _____

   _____

   How are they different? _____

   _____

2. Simulate the circuit of Figure 16-3. Compare the results of the simulation with the actual results obtained.

   How are the two results similar? _____

   _____

   How are they different? _____

   _____

Clippers, Clampers, and Detectors

Name: _____   Date: _____

Lab Section: _____ _____   Lab Instructor: _____
　　　　　　　　　　day　　　　　　　　time

Text Reference ◼

DC/AC Circuits and Electronics: Principles and Applications
Chapter 15: Wave Shaping and Generation

Materials Required ◼

Triple power supply (2 @ 0–20 volts DC; 1 @ 5 volts DC)
Oscilloscope
Function generator
1　each 1-kΩ, 10-kΩ, and 100-kΩ resistor
1　1N4001 or equivalent
1　each 1-μF, 10-μF, and 100-μF capacitor
2　uA741 integrated circuits

Introduction ◼

In this exercise, you will:

- Examine a positive and a negative clipper
- Examine a positive and a negative clamper circuit
- Examine a peak voltage detector.

Pre-Lab Activity Checklist

☐ Find the expected values in Table 17-1.
☐ Find the expected values in Table 17-2.
☐ Find the expected values in Table 17-3.
☐ Find the expected values in Table 17-4.
☐ Find the expected values in Table 17-5.
☐ Build the circuit of Figure 17-1.

Performance Checklist

☐ Pre-lab completed?   🛑 Instructor sign-off _____
☐ Table 17-1: Demonstrate measured V_{max} and V_{min}.
☐ Table 17-3: Demonstrate measured V_{max}, V_{min}, and DC.
☐ Table 17-5: Demonstrate measured DC output for a 4-V input.

Procedure 17-1

The Positive Clipper

1. Apply a sinusoidal wave input of 6-V_{pp}, 1-kHz with no DC offset to the circuit of Figure 17-1. Observe the output with an oscilloscope and sketch the output waveform in Figure 17-2.

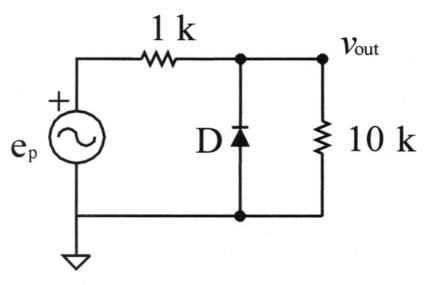

FIGURE 17-1 Positive Clipper Circuit

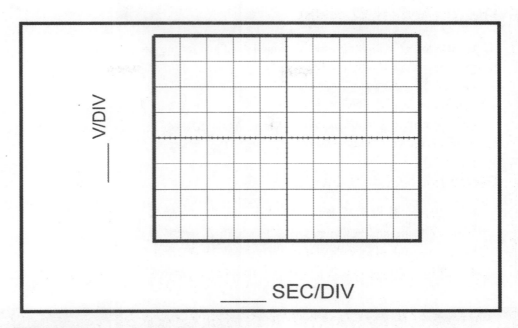

FIGURE 17-2 Positive Clipper Output Waveform

TABLE 17-1 Positive Clipper Circuit Data

Voltage Point	Expected Value	Measured Value
V_{max}		(STOP)
V_{min}		(STOP)

(STOP) Instructor sign-off of measured values in Table 17-1 _____

2. Measure the output voltage points, V_{min} and V_{max}; record the values in Table 17-1.

SAMPLE CALCULATIONS
_____

Show your sample calculations for:

$V_{max} =$

$V_{min} =$

Observations

1. What is the peak forward current through the diode for the circuit of Figure 17-1?

   $I_{\text{forward-peak}} =$ _____

2. What is the peak reverse voltage across the diode for the circuit of Figure 17-1?

   $V_{\text{reverse-peak}} =$ _____

3. What would the peak forward current be for the circuit of Figure 17-1 if the diode were germanium instead of silicon?

   $I_{\text{forward-peak}} =$ _____

Procedure 17-2

The Negative Clipper

1. Construct the circuit of Figure 17-3. Apply a sine wave input of 6 V_{pp}, 1-kHz frequency with no DC offset. Measure the output with an oscilloscope and draw the output waveform in Figure 17-4.

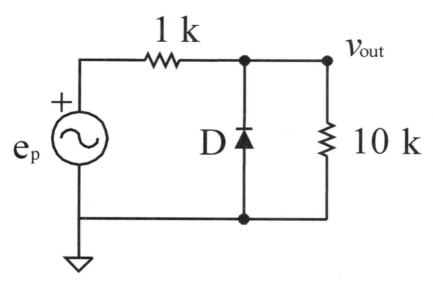

FIGURE 17-3 Negative Clipper Circuit

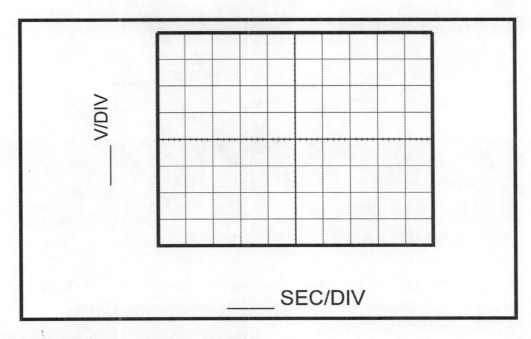

FIGURE 17-4 Negative Clipper Output Waveform

2. Measure the voltage points V_{min} and V_{max} on the output waveform and record these values in Table 17-2 in the measured values column.

TABLE 17-2 Negative Clipper Circuit Data

Voltage Point	Expected Value	Measured Value
V_{max}		
V_{min}		

SAMPLE CALCULATIONS

Show your sample pre-lab calculations for:

$V_{max} =$

$V_{min} =$

Observation ■

1. Which resistor in the circuit of Figure 17-3 determines the value of the peak reverse current?

Procedure 17-3

The Positive Clamper

1. Construct the circuit of Figure 17-5. Apply a sine wave input of 6 V_{pp}, 1-kHz frequency with no DC offset. Measure the output with an oscilloscope and draw the output waveform in Figure 17-6.

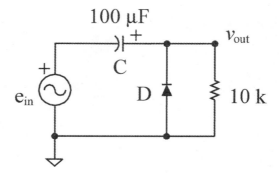

FIGURE 17-5 Positive Clamper Circuit

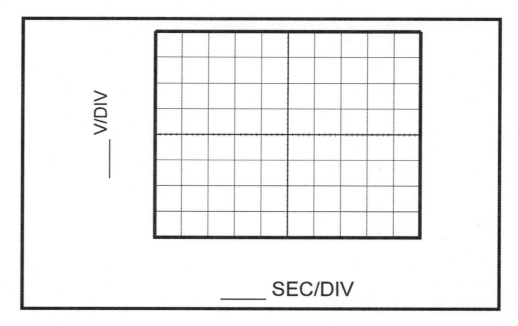

FIGURE 17-6 Positive Clamper Waveform

2. Measure the voltage points V_{min} and V_{max} on the output waveform and record these values in Table 17-3 in the measured values column and on the waveform of Figure 17-6.

TABLE 17-3 Positive Clamper Circuit Data

Voltage Point	Expected Value	Measured Value
V_{max}		(STOP)
V_{min}		(STOP)
DC		(STOP)

(STOP) Instructor sign-off of measured values in Table 17-3 _____

SAMPLE CALCULATIONS

Show your sample pre-lab calculations for:

$V_{max}=$

$V_{dc} =$

Observation ▪

1. If we assume that the RC time constant of the clamper of Figure 17-5 must be ten times that of the input signal period, what is the minimum frequency at which the clamper can operate?

$f_{min} =$ _____

Procedure 17-4

The Negative Clamper

1. Construct the circuit of Figure 17-7. Note that the diode and capacitor are reversed from the positive clamper.

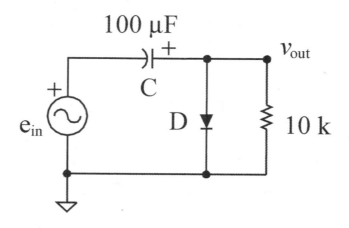

FIGURE 17-7 Negative Clamper Circuit

2. Apply a sine wave input of 6-V_{pp}, 1-kHz frequency with no DC offset. Measure the output with an oscilloscope and draw the output waveform in Figure 17-8.

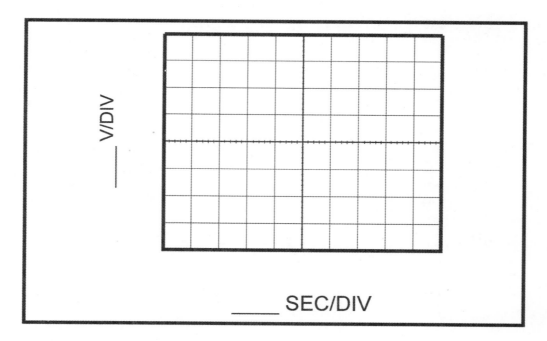

FIGURE 17-8 Negative Clamper Waveform

3. Measure the voltage points V_{min} and V_{max} on the output waveform and record these values in Table 17-4 in the measured values column and on the waveform of Figure 17-8.

TABLE 17-4 Negative Clamper Circuit Data

Voltage Point	Expected Value	Measured Value
V_{max}		
V_{min}		
DC		

SAMPLE CALCULATIONS

Show your sample pre-lab calculations for:

$V_{max} =$

$V_{dc} =$

Observation   ■

1. If the capacitor in Figure 17-7 were reduced to 1 µF, what effect would that have on the output waveform?

_____

_____

Synthesis

The Peak Detector

Select appropriate values for C and R_{load} for the peak-to-peak detector in Figure 17-9 based on the input signal and the op-amp characteristics. Record your selected values.

C = _____ R_{load} = _____

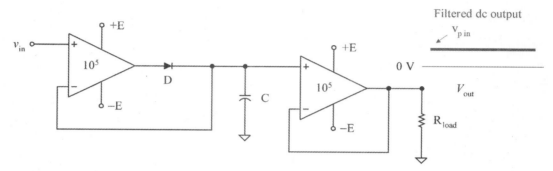

FIGURE 17-9   Active Peak-to-Peak Detector

Construct the circuit of Figure 17-9, then apply a sinusoidal input of 5-V_{pp}, 100-Hz frequency with no DC offset. Measure the output with an oscilloscope and sketch the output waveform in Figure 17-10. Vary the input amplitude as indicated in Table 17-5 and record the output voltages.

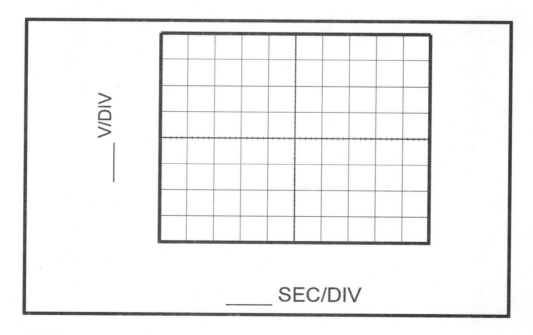

FIGURE 17-10   Detector Output Waveform

TABLE 17-5 Peak Voltage Detector Data

$V_{in\,pp}$	Expected DC Output	Measured DC Output
1 V		
2 V		
3 V		
4 V		
5 V		

STOP Instructor sign-off of measured value in Table 17-5 _____

SAMPLE CALCULATIONS

Show your sample calculations for:

$V_{dc\,(5\,V)} =$

Observation

1. If a 1-V_{dc} offset were added to AC input of the circuit of Figure 17-9 what effect would it have on the output voltage?

●*RL* Circuits

Name: _____   Date: _____

Lab Section: _____ _____   Lab Instructor: _____
              day   time

Text Reference ▪

DC/AC Circuits and Electronics: Principles and Applications
Chapter 16: Inductance and *RL* Circuits

Materials Required ▪

Triple power supply (2 @ 0–20 volts DC; 1 @ 5 volts DC)
Oscilloscope
Function generator
1   each 470-Ω and 10-kΩ resistor
1   each 22-mH, 33-mH, and 1-mH inductor

Introduction ▪

In this exercise, you will:

- Examine a series *RL* circuit
- Examine the voltage response of an inductor to a square wave input
- Examine how an inductor's reactance varies with the frequency of the signal applied.

Pre-Lab

Pre-Lab Activity Checklist

- ☐ Build the circuit of Figure 18-2.
- ☐ Find the expected values in Table 18-2.
- ☐ Find the expected values in Table 18-3.
- ☐ Find the expected values in Table 18-4.

Performance Checklist

- ☐ Pre-lab completed?   🛑 Instructor sign-off _____
- ☐ Table 18-2: Demonstrate V_L and V_R at 20 kHz.
- ☐ Table 18-3: Demonstrate V_L and V_R at 5 kHz.
- ☐ Table 18-4: Demonstrate measured τ for the 22-mH inductor.

Tutorial

Measuring the Time Constant, τ, with an Oscilloscope

To measure the time constant with the oscilloscope, the transient portion of an exponential waveform response is used. The procedure to measure the time constant, τ, is the same for *RL* circuits and for *RC* circuits. Figure 18-1 reviews this technique. First, select a convenient transient portion of the exponential curve to measure. The time constant is the same for all of the transient portions of the curve; therefore, choose a portion that is the most convenient and accurate to measure.

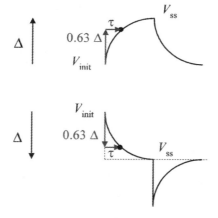
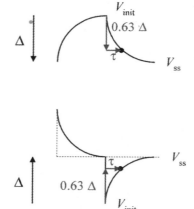

FIGURE 18-1   *RL* Transient Response Curves

☐ Select the transient exponential curve.

☐ Set up the time scale such that the signal fills up as much of the display as possible to obtain the most accurate measurement. (You may wish to take the V/DIV out of calibration so that the waveform fills the entire 8 vertical divisions).

☐ Measure the number of divisions between the steady state and the initial values (shown as Δ in Figure 18-1).

☐ Calculate 63% of Δ. (If you use 8 divisions, 63% is approximately 5 vertical divisions).

☐ Start at the initial value point and go up (or down) 63% of Δ.

☐ Go horizontally across the screen and count the number of divisions until the transient curve is intersected. You may increase the SEC/DIV setting to improve the accuracy.

☐ Using the oscilloscope's time scale setting, translate the number of horizontal divisions into time. That is one time constant, τ.

Procedure 18-1

Reactance of the Inductor

1. Measure the components listed in Table 18-1 and record their values. Measure the resistance and inductance of the inductors. Do not measure the inductance of the resistors.

TABLE 18-1 Component Measurements

Nominal Value	Measured Resistance	Measured Inductance
1 kΩ		
10 kΩ		
1 mH		
22 mH		
33 mH		

2. Set the AC input to 5 V$_p$ with no DC offset at 2 kHz and apply it to the circuit in Figure 18-2. Measure the AC output voltage for the inductor and resistor and record them in Table 18-2.

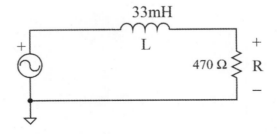

33mH

L

470 Ω R

FIGURE 18-2 Series *RL* Circuit

TABLE 18-2 Series *RL* Circuit Data

Frequency	Measured V_L	Actual V_L Change	Expected V_L Change	Measured V_R	Actual V_R Change	Expected V_R Change
250 Hz		↑ ↓ →	↑ ↓ →		↑ ↓ →	↑ ↓ →
500 Hz		↑ ↓ →	↑ ↓ →		↑ ↓ →	↑ ↓ →
1 kHz		↑ ↓ →	↑ ↓ →		↑ ↓ →	↑ ↓ →
2 kHz						
5 kHz		↑ ↓ →	↑ ↓ →		↑ ↓ →	↑ ↓ →
10 kHz		↑ ↓ →	↑ ↓ →		↑ ↓ →	↑ ↓ →
20 kHz	(STOP)	↑ ↓ →	↑ ↓ →	(STOP)	↑ ↓ →	↑ ↓ →

Legend: ↑ increase   ↓ decrease   → no change

(STOP) Instructor sign-off of measured values in Table 18-2 _____

3. Change the AC input voltage to the first value less than 2 k-Hz in Table 18-2. Indicate whether the voltage measured decreased, increased, or did not change compared to the 2 kHz input frequency.

4. Change the AC input voltage to each of the remaining values in Table 18-2. Measure the voltage across the inductor and the resistor and record the values measured in Table 18-2. Indicate whether the value measured decreased, increased, or stayed the same compared to the 2-kHz input voltage.

SAMPLE CALCULATIONS

Show your sample calculations made to determine the change in V_L between 2 kHz and 1 kHz:

Observations

1. What happened to the AC voltage across the inductor in the circuit of Figure 18-2 when the input frequency increased?

☐ Increased ☐ Decreased

Explain why. _____

_____

2. What happened to the AC voltage across the resistor in the circuit of Figure 18-2 when the input frequency increased?

☐ Increased ☐ Decreased

Explain why. _____

_____

Procedure 18-2

DC Voltage Response to an *RL* Circuit

1. Calculate the expected DC voltage across the inductor and resistor and record that value in Table 18-3. Use the measured resistance of the inductor from Table 18-1 to make these calculations.

TABLE 18-3 Series RL Circuit with DC Offset Data

Frequency	Expected $V_{L\,(dc)}$	Expected $V_{R\,(dc)}$	Measured $V_{L\,(dc)}$	Measured $V_{R\,(dc)}$
1 kHz				
2 kHz				
5 kHz			🛑	🛑
10 kHz				
20 kHz				

🛑 Instructor sign-off of measured values in Table 18-3 _____

2. Set the AC input to 5 V_p with 2 V_{dc} offset for the circuit of Figure 18-3. Measure the DC output voltage and record.

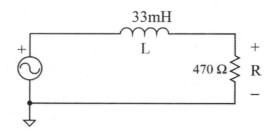

FIGURE 18-3 Series *RL* Circuit with DC Offset

3. Set the AC input frequency to each of the values listed in Table 18-3 and record the DC output voltage for each. Record the DC voltage measured at each value of the input frequencies in Table 18-3.

SAMPLE CALCULATIONS

Show your sample DC voltage calculations at 5 kHz for:

$V_L =$

$V_R =$

Observation

1. What happened to the DC voltage across the inductor in the circuit of Figure 18-2 when the input frequency increased?

   □ Did not change   □ Increased   □ Decreased

   Explain why. _____

   _____

Procedure 18-3

Square Wave Input and Measuring τ

1. Connect the circuit of Figure 18-4. Set the input square wave to 4 V_{pp} at 1 kHz with 0 V_{dc} offset.

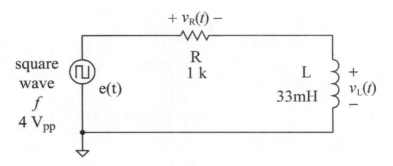

FIGURE 18-4 *RL* Circuit Square-Wave Response

2. Measure the output waveform on the inductor with Channel A of your oscilloscope. Connect Channel B to the input side of the resistor; depress the "inv channel B" switch and set the display of the oscilloscope to SUM A and B. In some cases, the V/DIV on both channels must be the same.

    ↘ **Note:**   The exact switch names may vary depending on the oscilloscope that you use; look for the equivalent wording. The display of SUM of A and B with Channel B inverted effectively subtracts Channel B from Channel A, giving the difference in voltage between the two probes, which is the voltage across the resistor.

3. Sketch the resultant waveforms for V_R and V_L in Figure 18-5.

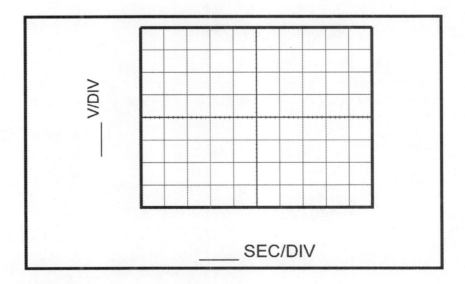

FIGURE 18-5 *RL* Circuit Output Waveform

4. Based on the measurements, determine the time constant of the circuit and record this value in Table 18-4.

5. Substitute the 22-mH inductor for the 33-mH inductor in the circuit of Figure 18-4. Measure the time constant of the circuit and record your result in Table 18-4.

TABLE 18-4   Square Wave *RL* Circuit Data

Inductor	τ Expected Value	τ Measured Value
33 mH		
22 mH		(STOP)

(STOP)   Instructor sign-off of measured values in Table 18-4 _____

SAMPLE CALCULATIONS

Show your sample calculations for:

$\tau =$

Observations ◼

1. Did the time constant of the circuit of Figure 18-4 increase or decrease when the 33-mH inductor was replaced by the 22-mH inductor?

    ☐ Increased ☐ Decreased

    What equation relates to this effect? _____

    _____

2. If the resistor in Figure 18-4 were doubled from its present value, what effect would that have on the time constant of the circuit?

    ☐ It would increase ☐ It would decrease

    Explain why. _____

    _____

Synthesis

Altering the Time Constant of an Inductive Circuit

Using the circuit of Figure 18-4, select any combination of the following components to obtain the longest possible time constant.

- 22-mH inductor
- 33-mH inductor
- 1-kΩ resistor
- 10-kΩ resistor

Calculate the expected value of τ for the components you selected, then build the circuit and measure τ. Record the values in Table 18-5.

TABLE 18-5 Maximized Value of τ	
τ **Expected Value**	τ **Measured Value**

Transformers and Half-Wave Rectifier Circuits

Name: _____ Date: _____

Lab Section: _____ _____ Lab Instructor: _____
　　　　　　　　　day　　　　　　　　time

Text Reference ◼

DC/AC Circuits and Electronics: Principles and Applications
Chapter 18: Power Supply Applications

Materials Required ◼

　　Triple power supply (2 @ 0–20 volts DC; 1 @ 5 volts DC)

　1　120 to 12.6-V transformer

　4　1N4001 diodes or equivalent

　1　1-kΩ resistor

　1　470-μF electrolytic capacitor

Introduction ◼

In this exercise, you will:

- ◼ Examine the characteristics of the transformer
- ◼ Examine triggering the oscilloscope
- ◼ Examine a filtered and unfiltered half-wave rectifier

Pre-Lab Activity Checklist ■

☐ Find the expected values in Table 19-1 through Table 19-5.
☐ Find the expected values in Table 19-7 through Table 19-10.

Performance Checklist ■

☐ Pre-lab completed?

☐ Table 19-1: Demonstrate measured V_{rms}.

☐ Table 19-3: Demonstrate measured T.

☐ Table 19-9: Demonstrate measured V_{ripple}.

🛑 Instructor sign-off _____

Tutorial

Oscilloscope Trigger Controls

Triggering controls and corresponding display controls provide flexibility in properly displaying an observed signal.

A trigger signal is required to start the trace of an observed signal on CH1 or CH2. If a trigger signal is not present, the oscilloscope will not display the signal to be observed. The observed signals on CH1 and CH2 start their traces based upon the oscilloscope's trigger settings.

There are three sources for trigger signals. Those sources are:

☐ **INT**   (Internal)   Triggers on CH1 or CH2 signal (depends on CH1-VERT-CH2 setting)
☐ **LINE**   (Line)   Triggers on the commercial LINE voltage (for 60 Hz and harmonics)
☐ **EXT**   (External)   Triggers on external signal applied to the EXT jack

If the trigger is set to **INTERNAL**, the observed signals on CH1 and CH2 start their trace based upon the oscilloscope setting of the CH1-VERT-CH2 control. These settings are:

☐ **CH1**   Both channels trigger on CH1 signal (CH1 is the reference channel)
☐ **CH2**   Both channels trigger on CH2 signal (CH2 is the reference channel)
☐ **VERT**   (Vertical)   CH1 triggers on CH1 signal and CH2 triggers on CH2 signal

The trigger **SLOPE** control determines the start of the trace based upon the *slope of the triggering* signal. These settings are:

☐ **Positive**   ⌐   Slope setting starts trace when triggering signal's slope goes positive
☐ **Negative**   ⌐   Slope setting starts trace when triggering signal's slope goes negative

The trigger **LEVEL** control adjusts the starting voltage level of the observed signal. For example, one would use this control to start the observed sine wave signal at the 0-V level for easier observation.

Procedure 19-1

Transformer Characteristics

1. Measure and record the transformer secondary RMS voltages in Table 19-1. Refer to Figure 19-1.

TABLE 19-1   Transformer Secondary RMS Voltages

Secondary	Expected Value	Measured Value	% Error
Top to Bottom	15 V$_{rms}$	(STOP)	
Top to Center			
Bottom to Center			

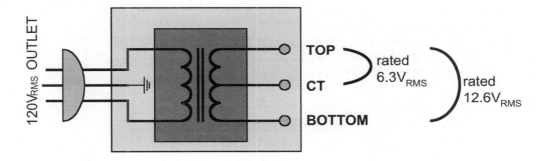

FIGURE 19-1   Rated 12.6-V$_{rms}$ Center-Tapped Transformer

▣ **Note:**   Use a 120 to 12.6-V center-tapped (CT) transformer for this experiment. The transformer primary should be plugged into a 120 V$_{rms}$ electrical outlet. The full secondary rating of 12.6 V$_{rms}$ is based upon delivering its secondary current rating (typically 1–2 A$_{rms}$). If the transformer is unloaded or lightly loaded, its output will be higher (typically 15 V$_{rms}$). Use 15 V$_{rms}$ as the secondary voltage in calculating estimated values when the transformer is lightly loaded (significantly less than 1 A).

2. Measure and record the transformer secondary DC voltages in Table 19-2.

TABLE 19-2   Transformer Secondary DC Voltages

Secondary	Expected Value	Measured Value	% Error
Top to Bottom			
Top to Center			
Bottom to Center			

3. Using the oscilloscope, measure the voltage across the full secondary as shown in Figure 19-2. Set the oscilloscope controls to accurately display the secondary voltage and period. Sketch the voltage waveform on Figure 19-3.

Set up the oscilloscope:

☐ CH1 ☐ DC coupling ☐ INT trigger on CH1

☐ Ground trace centered ☐ SLOPE control is positive ⌐

☐ LEVEL adjusted to start signal display at 0 V

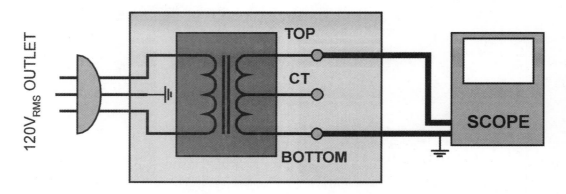

FIGURE 19-2 Oscilloscope Measurements of Full Transformer Secondary

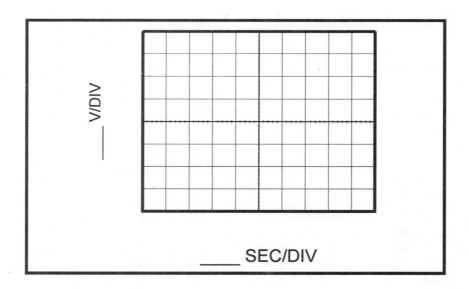

FIGURE 19-3 Oscilloscope Measurement of Full Secondary

4. Observe and record the peak-to-peak voltage, peak voltage, period, and frequency in Table 19-3.

TABLE 19-3 Transformer Secondary Voltage

Full Secondary	Expected Value	Measured Value	% Error
V_{pp}			
V_p			
T		⛔	
f			

⛔ Instructor sign-off of measured values in Table 19-3 _____

SAMPLE CALCULATIONS _____

Show your pre-lab calculations for:

$V_p =$

---

5. Without changing the oscilloscope settings, measure the voltage from the TOP to the CENTER TAP of the secondary as shown in Figure 19-4. Sketch the voltage waveform on Figure 19-5.

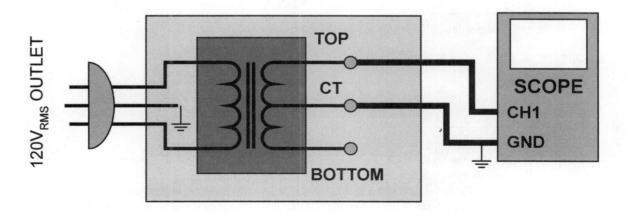

FIGURE 19-4 Oscilloscope Measurements of Top Half of Transformer Secondary (*CT connected to oscilloscope ground*)

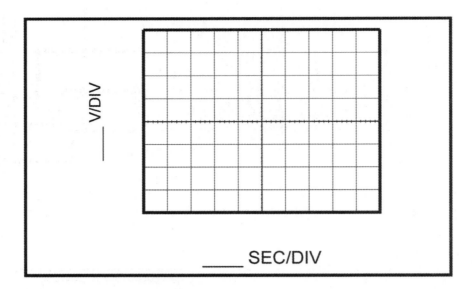

FIGURE 19-5 Oscilloscope Measurement of Top Half Secondary

6. Observe and record the peak-to-peak voltage, peak voltage, period, and frequency in Table 19-4.

TABLE 19-4 Top to Center-Tap Secondary Voltage

Top Half of Secondary	Expected Value	Measured Value	% Error
V_{pp}			
V_p			
T			
f			

SAMPLE CALCULATIONS

Show your pre-lab calculations for the top half of the secondary:

$V_p =$

7. Set the oscilloscope mode to display both CH1 and CH2. Center the CH1 GND line on the top half of the display and the CH2 GND line on the bottom half of the display. Re-observe the CH1 signal and adjust the sensitivity to observe the entire waveform. Set the CH2 sensitivity to match CH1. Be sure the CH2 INV button is not depressed. Again, use the horizontal position and trigger level controls to anchor this trace starting at the 0-V line. Connect CH2 to the **bottom** of the secondary (see Figure 19-6). Sketch the CH1 and CH2 waveforms to scale on Figure 19-7. Measure and record the **bottom** half signal data in Table 19-5.

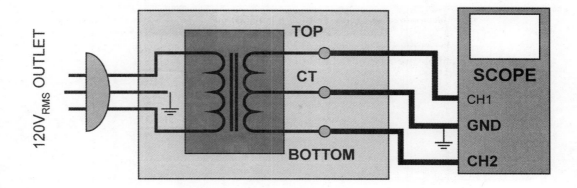

FIGURE 19-6 Oscilloscope Measurements of Bottom Half of Secondary (*CT connected to oscilloscope ground*)

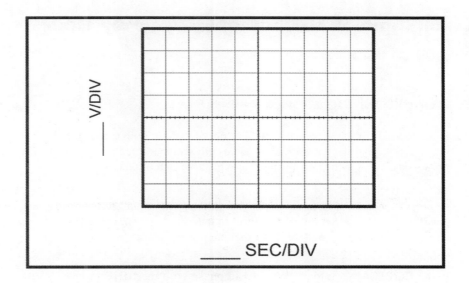

FIGURE 19-7 Oscilloscope Measurement of Top *and* Bottom Halves of Secondary

8. Observe and record the peak-to-peak voltage, peak voltage, and period in Table 19-5.

TABLE 19-5 Bottom to Center-Tap Secondary Voltage			
Bottom Half of Secondary	**Expected Value**	**Measured Value**	**% Error**
V_{pp}			
V_{p}			
T (scope)			
f (DMM)			

SAMPLE CALCULATIONS

Show your sample pre-lab calculations for:

$f =$

Tutorial
Understanding Oscilloscope Triggering Modes

Phase Relationship Observations ◼

1. Is the top half of the secondary "in" or "out of" phase with the bottom half of the secondary? To determine this, look at the waveforms on the oscilloscope and see if v_{bottom} is positive when v_{top} is positive (in phase) or if v_{bottom} is negative when v_{top} is positive (out of phase).

    ☐ In phase ☐ Out of phase

2. Is the TOP secondary terminal in phase with the LINE input? To determine this, use line triggering with a positive slope (⌐). Set the trigger level to the middle (0 V). If the TOP terminal display has the positive half-cycle first (left side of display), then they are in phase.

    ☐ Yes ☐ No

3. Is the BOTTOM secondary terminal in phase with the LINE input? Use the technique above on the BOTTOM terminal.

    ☐ Yes ☐ No

4. Remove both oscilloscope probes, and then reconnect CH1 across the full secondary (see Figure 19-2). Return the trigger to **INT**ernal **CH1** with a positive slope. The observed trace on CH1 has been triggering off of itself, thus

   ◼ CH1 has been positive-going relative to itself (in phase with itself).
   ◼ CH2 (now *not* connected) starts its trace relative to the CH1 signal.

Observations of Other Triggering Options ◼

(while applying a signal to only CH1)

5. **NO** Triggering Signal. Change the trigger to **INT**ernal **CH2**.

    What happened? _____

    Why? _____

6. **NO** Triggering Signal. Change the trigger to **EXT**ernal.

    What happened? _____

    Why? _____

7. **LINE** (commercial power) Triggering. Change the trigger to **LINE**.

What happened? _____

Why? _____

8. **LINE LINE** Triggering with **NEGATIVE**-Going Slope Trigger. Change the trigger **SLOPE** to **NEGATIVE** ($\diagdown\_$).

What happened? _____

Why? _____

Return the trigger to **INT**ernal **CH1** with a **POSITIVE**-going **SLOPE**. The signal should return to its original display.

Observations ■

1. Did the transformer secondary produce only AC or was there DC voltage present?

_____

_____

2. Why is it *not* safe to measure only AC or only DC?

_____

_____

3. What is the frequency of the AC signal on the secondary of the transformer?

_____

_____

4. Why would you want to choose LINE triggering instead of INTernal triggering to observe commercial voltages?

_____

_____

Procedure 19-2

Lightly-Loaded, Positive Half-Wave Rectifier Power Supply

1. Measure and record in Table 19-6 the values of load R_L and filter capacitor C of Figure 19-8.
2. Construct the unfiltered rectifier circuit of Figure 19-8 (SW open), connecting the oscilloscope as shown. Set up the oscilloscope:

☐ CH1   ☐ DC coupling   ☐ Ground trace centered

☐ INT trigger on CH1   ☐ SLOPE control is positive $\diagup$

☐ LEVEL adjusted to start signal display at 0 V

TABLE 19-6 Component Values

Component	Expected Value	Measured Value	% Error
R_L	1.0 kΩ		
C	470 μF		

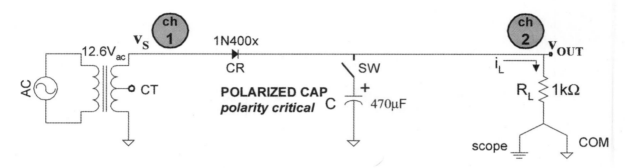

FIGURE 19-8 Positive Half-Wave Power Supply (15-V_{rms}, lightly-loaded secondary)

3. Set up the oscilloscope to accurately observe the transformer secondary voltage V_S on CH1 and sketch it on Figure 19-9.

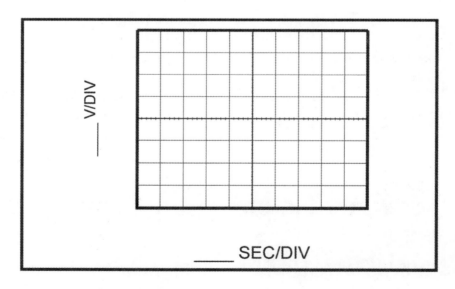

FIGURE 19-9 Power Supply Waveforms, Half-Wave

Legend: ••••••• Transformer Secondary Voltage (CH1)
– – – – *Unfiltered* Power Supply Output Voltage (CH2)
———— *Filtered* Power Supply Output Voltage (with cap)

⬜ **Note:** You will be sketching three graphs on this figure; follow the legend or use different colors.

4. Using this oscilloscope display, measure and record V_S peak voltage (relative to ground) and period in Table 19-7.

TABLE 19-7 Transformer Secondary Voltage, Half-Wave Rectified (*use DMM)

V_S	Expected Value	Measured Value	% Error
V_S relative to COM			
T			

5. Sketch the unfiltered power supply output voltage V_{out} on CH2 on Figure 19-9.
6. Using the oscilloscope display, measure and record V_{out} peak voltage and period in Table 19-8.

TABLE 19-8 Unfiltered Output Voltage, Half-Wave Rectified

V_{out}	Expected Value	Measured Value	% Error
V_p			
V_{dc}			
T			

7. Using the DMM, measure the power supply's DC output voltage.
8. Close switch SW to create a filtered power supply.
9. Observe the filtered power supply output voltage V_{out} on CH2 and sketch it on Figure 19-9.
10. Using this oscilloscope display, measure and record V_{out} peak voltage and period in Table 19-9.

TABLE 19-9 Filtered Output Voltage, Half-Wave Rectified

V_{out}	Expected Value	Measured Value	% Error
V_p			
$V_{pp\ ripple}$		STOP	
V_{dc}			
$V_{rms\ ripple}$			
T			

STOP Instructor sign-off of measured values in Table 19-9 _____

11. Using the DMM, measure the power supply's DC output voltage.

12. Using the DMM, measure the power supply's AC output voltage (RMS ripple voltage)

13. Observe the ripple voltage on the oscilloscope on CH2. Change the oscilloscope to AC coupling, center the GND line, and change CH2 voltage sensitivity to accurately observe the power supply ripple voltage. Measure and record the peak-to-peak ripple voltage and the period in Table 19-9.

SAMPLE CALCULATIONS

Show your pre-lab calculations for:

$V_S =$

SAMPLE CALCULATIONS

Show your sample pre-lab calculations for:

$V_p =$

SAMPLE CALCULATIONS

Show your pre-lab calculations for:

$V_{pp\ ripple} =$

---

14. Open the switch to produce an unfiltered power supply again. Change the oscilloscope to observe the diode voltage using the oscilloscope CH1-CH2 feature.
    Set up the oscilloscope:
    ☐ BOTH ☐ ADD ☐ INV CH2 ☐ DC Coupling ☐ CH1 & CH2 same sensitivities
    ☐ Ground Line on the top horizontal oscilloscope display line

15. Observe and sketch the diode voltage waveform on Figure 19-10. Note the maximum reverse voltage experienced by the diode and record it in Table 19-10.

16. Close switch SW to produce a filtered power supply again. Observe the diode voltage. Observe and sketch the diode voltage waveform on Figure 19-10. Note the maximum reverse voltage experienced by the diode and record it in Table 19-10.

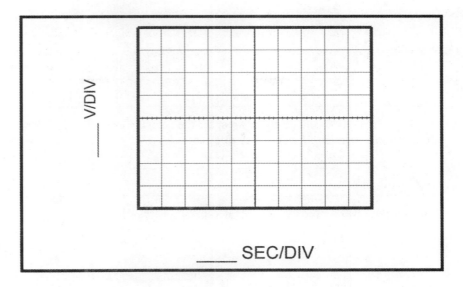

FIGURE 19-10 Diode Voltage Waveforms, Half-Wave

Legend: – – – – *Unfiltered* Power Supply Output Voltage (CH2)
———— *Filtered* Power Supply Output Voltage (with cap)

TABLE 19-10 Diode Reverse Voltage Measurements, Half-Wave

	Expected Peak Reverse Diode Voltage	Measured Peak Reverse Diode Voltage	% Error
Unfiltered			
Filtered			

SAMPLE CALCULATIONS

Show your sample pre-lab calculations for:

$V_{p \, (rev) \, unfiltered} =$

Observations

1. What purpose does the rectifier diode serve in this circuit?

_____

_____

2. Based upon the diode voltage curve, what are the peak inverse voltages experienced by the unfiltered and filtered power supplies? Why are they different?

_____

_____

3. What purpose was served by placing the capacitor across the load?

_____

_____

4. What is the percent ripple of this circuit?

_____

_____

Full-Wave Rectifier Circuits

Name: _____ Date: _____

Lab Section: _____ _____ Lab Instructor: _____
day   time

Text Reference

DC/AC Circuits and Electronics: Principles and Applications
Chapter 18: Power Supply Applications

Materials Required

Triple power supply (2 @ 0–20 volts DC; 1 @ 5 volts DC)
Oscilloscope
1   120 to 12.6-V transformer
4   1N4001 diodes or equivalent
1   1-kΩ resistor
1   470-µF electrolytic capacitor

Introduction

In this exercise, you will:

- Examine a filtered and unfiltered center-tapped full-wave rectifier
- Examine a filtered and unfiltered full-wave bridge rectifier.

Pre-Lab Activity Checklist ◼

☐ Find the expected values for Tables 20-1 through 20-6.
☐ Build the circuit of Figure 20-1.

Performance Checklist ◼

☐ Pre-lab completed?   🛑 Instructor sign-off _____
☐ Table 20-2: Demonstrate measured V_{dc}.
☐ Table 20-3: Demonstrate measured f.
☐ Table 20-6: Demonstrate measured V_p

Procedure 20-1

Lightly-Loaded, Positive Full-Wave Center-Tapped Rectifier

1. Construct the unfiltered rectifier circuit of Figure 20-1 (SW open), connecting the oscilloscope as shown.

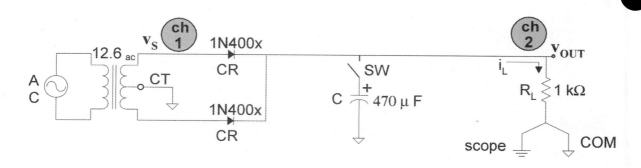

FIGURE 20-1 Positive Full-Wave, Center-Tapped Power Supply (lightly-loaded secondary)

Set up the oscilloscope:

☐ CH1   ☐ DC coupling   ☐ Ground trace centered
☐ INT trigger on CH1   ☐ SLOPE control is positive ⌐
☐ LEVEL adjusted to start signal display at 0 V

2. Set up the oscilloscope to accurately observe the transformer secondary voltage V_S on CH1 and sketch it on Figure 20-2.

   ⊐ Note: You will be sketching three graphs on this figure; follow the legend or use different colors.

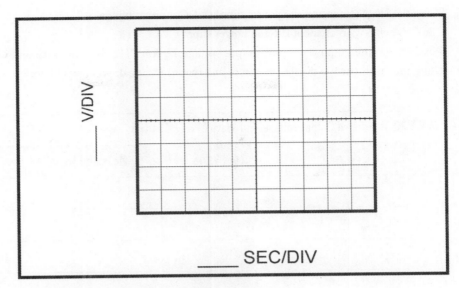

FIGURE 20-2 Power Supply Waveforms, Full-Wave CT

Legend: ●●●●●●● Transformer Secondary Voltage (CHI)
        – – – – *Unfiltered* Power Supply Output Voltage (CH2)
        ———— *Filtered* Power Supply Output Voltage (with cap)

3. Using this oscilloscope display, measure and record V_S peak voltage (relative to ground) and period in Table 20-1.

TABLE 20-1 Transformer Secondary Voltage, Full-Wave CT

V_s	Expected Value	Measured Value	% Error
V_S relative to COM			
T			

4. Observe the unfiltered power supply output voltage V_{out} on CH2 and sketch it on Figure 20-2.
5. Using the oscilloscope display, measure and record V_{out} peak voltage and period in Table 20-2.

TABLE 20-2 Unfiltered Output Voltage, Full-Wave CT

V_{out}	Expected Value	Measured Value	% Error
V_p			
V_{dc}		🛑	
T			

🛑 Instructor sign-off of measured values in Table 20-2 _____

6. Using the DMM, measure the power supply's DC output voltage.
7. Close switch SW to create a filtered power supply.
8. Observe the filtered power supply output voltage V_{out} on CH2; sketch it also on Figure 20-2.
9. Using this oscilloscope display, measure and record V_{out} peak voltage and period in Table 20-3.

TABLE 20-3   Filtered Output Voltage, Full-Wave CT

V_{out}	Expected Value	Measured Value	% Error
V_p			
$V_{pp\ ripple}$			
V_{dc}			
$V_{rms\ ripple}$			
T		STOP	

STOP   Instructor sign-off of measured values in Table 20-3 _____

10. Using the DMM, measure the power supply's DC output voltage.
11. Using the DMM, measure the power supply's AC output voltage (RMS of the ripple voltage).
12. Observe the ripple voltage on the oscilloscope accurately on CH2. Change the oscilloscope to AC coupling, center the GND line, and change CH2 voltage sensitivity to observe accurately the power supply ripple voltage. Accurately measure and record the peak-to-peak ripple voltage and the period in Table 20-3.

SAMPLE CALCULATIONS

Show your pre-lab calculations for:

$V_S =$

SAMPLE CALCULATIONS

Show your pre-lab calculations for:

$V_P =$

SAMPLE CALCULATIONS

Show your pre-lab calculations for:

$V_p =$

$V_{rms\ ripple} =$

Observations ■

1. What is the frequency of the output voltage? _____

2. What is the ripple voltage frequency? _____

3. What is the percent change in the *unfiltered* DC output voltage compared to the half-wave rectifier of Exercise 19?

    _____

4. What is the percent change in the *filtered* DC output voltage compared to the half-wave rectifier of Exercise 19?

    _____

5. What is the percent ripple of this loaded circuit? _____
6. Is the full-wave rectifier an improvement over the half-wave circuit of Exercise 19?

    ☐ Yes ☐ No

Procedure 20-2

Lightly-Loaded, Positive Full-Wave Bridge Rectifier

1. Construct the unfiltered rectifier circuit of Figure 20-3 (SW open), connecting the oscilloscope as shown. Be sure to remove the common connection to the center tap of the transformer.

    Set up the oscilloscope:

    ☐ CH1   ☐ DC coupling   ☐ Ground trace centered
    ☐ INT trigger on CH1   ☐ SLOPE control is positive ⌐
    ☐ LEVEL adjusted to start signal display at 0 V

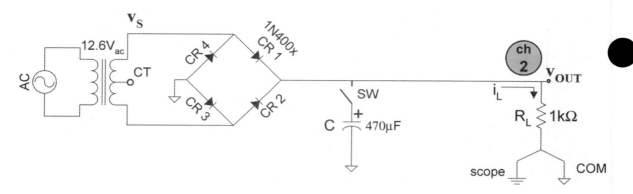

FIGURE 20-3 Positive Full-Wave Bridge (15-V$_{rms}$, lightly-loaded secondary)

2. Set up the oscilloscope to accurately observe the transformer secondary voltage V_S on CH1 and sketch it on Figure 20-4.

   Ↄ Note: You will be sketching three graphs on this figure; follow the legend or use different colors.

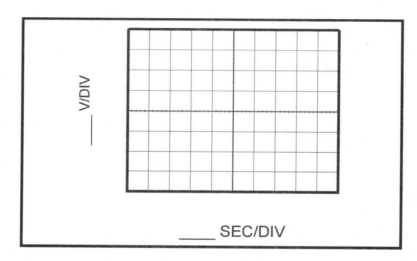

FIGURE 20-4 Power Supply Waveforms, Full-Wave Bridge

Legend: •••••••   Transformer Secondary Voltage (CH1)
          – – – –   *Unfiltered* Power Supply Output Voltage (CH2)
          ————   *Filtered* Power Supply Output Voltage (with cap)

3. Using this oscilloscope display, measure and record V_S peak voltage (relative to common) and period in Table 20-4.

TABLE 20-4	Transformer Secondary Voltage, Full-Wave Bridge		
V_s	**Expected Value**	**Measured Value**	**% Error**
V_S relative to COM			
T			

4. Observe the unfiltered power supply output voltage V_{out} on CH2; sketch it also on Figure 20-4.
5. Using the oscilloscope display, measure and record V_{out} peak voltage and period in Table 20-5.

TABLE 20-5 Unfiltered Output Voltage, Full-Wave Bridge

V_{out}	Expected Value	Measured Value	% Error
V_p			
V_{dc}			
T			

6. Using the DMM, measure the power supply's DC output voltage.
7. Close switch SW to create a filtered power supply.
8. Observe the filtered power supply output voltage V_{out} on CH2; sketch it also on Figure 20-4.
9. Using this oscilloscope display, measure and record V_{out} peak voltage and period in Table 20-6.

TABLE 20-6 Filtered Output Voltage Power Supply, Full-Wave Bridge

V_{out}	Expected Value	Measured Value	% Error
V_p		🛑	
$V_{pp\ ripple}$			
V_{dc}			
$V_{rms\ ripple}$			
T			

🛑 Instructor sign-off of measured values in Table 20-6 _____

10. Using the DMM, measure the power supply's DC output voltage.
11. Using the DMM, measure the power supply's AC output voltage (RMS of the ripple voltage).
12. Observe the ripple voltage on the oscilloscope accurately on CH2. Change the oscilloscope to AC coupling, center the GND line, and change CH2 voltage sensitivity to observe accurately the power supply ripple voltage. Accurately measure and record the peak-to-peak ripple voltage and the period in Table 20-6.

SAMPLE CALCULATIONS

Show your pre-lab calculations for:

$V_S =$

SAMPLE CALCULATIONS

Show your sample pre-lab calculations for:

$V_p =$

SAMPLE CALCULATIONS

Show your sample pre-lab calculations for:

$V_{pp} =$

Observations ■

1. What is the frequency of the output voltage? _____

2. What is the frequency of the ripple voltage? _____

3. What is the percent ripple of this loaded circuit? _____

4. Is the full-wave rectifier an improvement over the other two rectifier circuits?

   ☐ Yes   ☐ No

5. List the advantages and disadvantages of the full-wave rectifier over the other two rectifier circuits.

_____

_____

_____

_____

_____

_____

The Capacitor Input Filter and Zener Regulator

Name: _____ Date: _____

Lab Section: _____ _____ Lab Instructor: _____
            day   time

Text Reference

DC/AC Circuits and Electronics: Principles and Applications
Chapter 18: Power Supply Applications

Materials Required

Triple power supply (2 @ 0–20 volts DC; 1 @ 5 volts DC)
1   120 to 12.6-V transformer with center-tapped secondary
1   each 180-Ω, 330-Ω, 2.2-kΩ resistor
4   1N4001 diodes or equivalent
1   each 10-μF, 47-μF, 100-μF electrolytic capacitor
1   1N753 zener diode

Introduction

In this exercise, you will:

- Examine a bridge rectifier with a capacitor input filter
- Examine the effect that different filter capacitor values have on output ripple voltage
- Examine a zener diode regulator
- Examine the effect of varying the current-limiting resistor in a zener regulator
- Examine the effect of varying the load resistance on voltage regulation.

Pre-Lab Activity Checklist

☐ Find the expected values for Table 21-1 and include sample calculations.

☐ Find the expected values for Tables 21-3 through 21-5 and include sample calculations.

☐ Find the expected values for Table 21-7 and include sample calculations.

☐ Find the expected values for Table 21-9 and include sample calculations.

☐ Build the circuit of Figure 21-3.

Performance Checklist

☐ Pre-lab completed?   🛑 Instructor sign-off _____

☐ Table 21-3: Demonstrate measured $V_{out\,p}$.

☐ Table 21-4: Demonstrate measured $V_{out\,pp\,ripple}$.

☐ Table 21-7: Demonstrate measured $V_{out\,dc}$ and $V_{out\,pp\,ripple}$.

Procedure 21-1

Constructing a Power Supply: The Transformer

1. With the DMM measure and record the transformer full secondary voltage of Figure 21-1 in Table 21-1.

   ⤷ **Note:** this transformer output voltage is about 15 V_{rms} when lightly loaded.

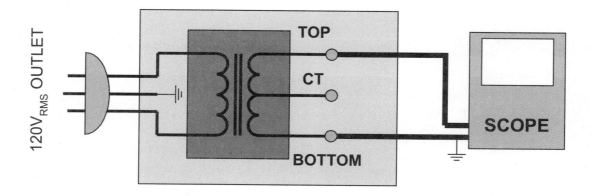

FIGURE 21-1 Transformer Circuit (rated 12.6 V_{ac}; 15 V_{ac} lightly loaded)

TABLE 21-1 Transformer Secondary Voltages (unloaded)

Quantity	Expected Value	Measured Value	% Error
$V_{\text{sec rms}}$	15 V_{rms}		
f			
$V_{\text{sec peak}}$			
T			

SAMPLE CALCULATIONS

Show your pre-lab calculations for:

$f =$_

$V_{\text{sec p}} =$_

$T =$_

2. Observe the full secondary voltage on CH1 of the oscilloscope using these settings: DC, CH1 INT trigger, centered GND line, and appropriate sensitivities.

3. Measure and record in Table 21-1 the secondary peak voltage V_{sec} and period T. Calculate the frequency.

4. Sketch the observed waveform in Figure 21-2. The secondary peak voltage is the key starting value in power supply circuit analysis.

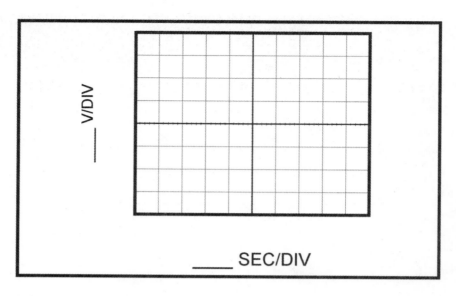

FIGURE 21-2 Transformer Secondary Waveform

Observations

1. What are the major functions of the transformer?

_____

_____

_____

2. What is the key voltage specification needed from the secondary of the transformer to begin power supply voltage calculations?

_____

_____

_____

Procedure 21-2

Constructing a Power Supply: Bridge Rectifier

1. Measure and record component values in Table 21-2.

TABLE 21-2 Component Measurements for Figure 21-3

Component	Nominal Value	Measured Value	% Error
R_L	2.2 kΩ		
C	100 µF		

2. Switch off the transformer supply and connect the secondary to the circuit of Figure 21-3 previously constructed. The switch (SW) is to be opened to produce an unfiltered output.

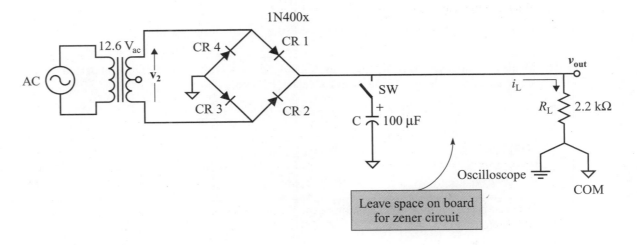

FIGURE 21-3 Full-Wave Rectifier Circuit (SW open, unfiltered, *pre-lab*)

 WARNING Connect your diodes correctly. If one diode is reversed, it may destroy two diodes. For example, if CR4 is reversed, CR3 and C4 are both forward biased on the positive half-cycle with a resistance of about one ohm. (I @ 20 V/1 W = 20 A)

Remember to keep your face away from a live circuit to prevent injury. Electrolytic capacitors can explode and diodes can vaporize. At the first sign of trouble, disconnect the power. Do not sniff out a hot smell; a capacitor or diode may blow up in your face!

3. Check your construction; one misconnection could damage the components. Consider having someone else check your circuit (and you check theirs). Once everything is correct, switch on the transformer.

4. Observe the full-wave rectified output voltage waveform with an oscilloscope and sketch it in Figure 21-4.

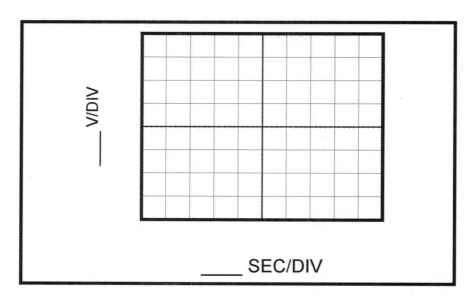

FIGURE 21-4   Rectified Full-Wave Output Voltage (without filtering)

5. Is the shape of the output voltage displayed a fully rectified sine wave?

□ Yes □ No

6. Measure the peak output voltage, peak-to-peak voltage, minimum voltage, and period with the oscilloscope and record them in Table 21-3. Calculate the output frequency.

TABLE 21-3   Rectified Full-Wave Output Voltage (without filtering)

Quantity	Expected Value	Measured Value	% Error
$V_{out\ p}$		🛑	
$V_{out\ pp}$			
$V_{out\ min}$			
T			
$V_{out\ dc}$			
f			

🛑 Instructor sign-off of measured values in Table 21-3 _____

7. Is the peak of the output voltage approximately 1.4 V less than the transformer secondary voltage?

☐ Yes ☐ No

8. Measure the DC output voltage with the DMM and record it in Table 21-3.

SAMPLE CALCULATIONS

Show your pre-lab calculations for:

$f =$

$V_{out\ p} =$

$T =$

Observations

1. What is the function of the rectifier circuit in the power supply?

_____

_____

2. How is the peak voltage out of the rectifier related to its input?

_____

_____

Procedure 21-3

Constructing a Power Supply: Filtering

1. Remove the power to the circuit and then close the switch (SW) to connect the 100-μF capacitor. See Figure 21-5.

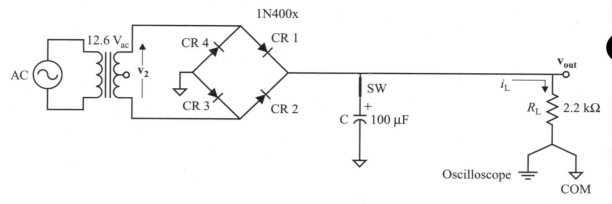

FIGURE 21-5 Filtered Full-Wave Rectifier (SW closed)

2. Read the DC voltage rating on your capacitor. _____ Is it sufficient?   ☐ Yes ☐ No
   If not, find a capacitor with a higher voltage rating.

 WARNING! Connect your polarized electrolytic capacitor correctly. If connected backward or if the DC voltage rating is exceeded, it may explode.

3. Keep your face away from the circuit. Power up the circuit. If you sense heat or your capacitor begins to swell, disconnect the power to the circuit immediately.

4. Use the oscilloscope to observe the output voltage waveform and sketch it in Figure 21-6.

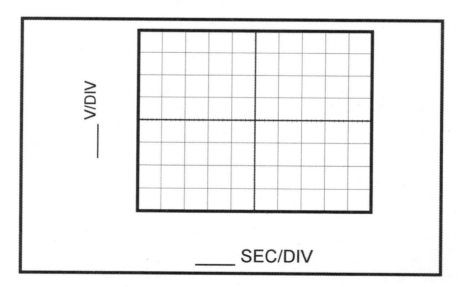

FIGURE 21-6 Filtered Full-Wave Output Voltage

5. Open and close the switch to observe the capacitor's effect on the output voltage.

6. Re-close the switch and measure the peak output voltage, the peak-to-peak ripple voltage, minimum voltage, and period of the ripple voltage with the oscilloscope. Record them in Table 21-4. Calculate the frequency of the ripple voltage.

TABLE 21-4 Filtered Full-Wave Output Voltage

Quantity	Expected Value	Measured Value	% Error
$V_{\text{out p}}$			
$V_{\text{out pp ripple}}{}^{*}$		(STOP)	
$V_{\text{out min}}$			
T			
$V_{\text{out dc}}$			
f			

*Use lightly-loaded linear approximation for small peak-to-peak ripple voltage (that is, less than 10% ripple or peak-to-peak ripple voltage less than 30% of peak voltage).

(STOP) Instructor sign-off of measured values in Table 21-4 _____

7. Is the filtered peak output voltage the same as it was in the unfiltered case?   ☐ Yes   ☐ No
8. Measure the DC output voltage and record it in Table 21-4.

SAMPLE CALCULATIONS

Show your pre-lab calculations for:

$f =$

$V_{\text{out p}} =$

$V_{\text{out min}} =$

$V_{\text{out pp ripple}}{}^{*} =$

*Use lightly-loaded linear approximation for small peak-to-peak ripple voltage (that is, less than 10% ripple or peak-to-peak ripple voltage less than 30% of peak voltage).

Approximate peak-to-peak ripple: $V_{pp} \cong \dfrac{I \cdot T}{C} = \dfrac{I_{dc} \cdot T}{C} \cong \dfrac{I_p \cdot T}{C} \cong \dfrac{I_p}{fC}$

where

T = period (16.6 ms for 60-Hz half-wave, 8.3 ms for 120-Hz full-wave)

f = frequency of rectified waveform in Hz

C = capacitance of filter capacitor

I = Load DC current as seen by capacitor

$\cong$ Load peak current for an initial, worst case approximation

9. Measure the DC output voltage $V_{out\,dc}$ and its frequency using the DMM and record it in Table 21-4. Note all the voltages in Table 21-4. What observations can you make?

_____

_____

10. Change the oscilloscope to the AC coupling and increase the sensitivity to accurately observe the AC ripple voltage. Draw a sketch of the AC ripple voltage in Figure 21-7. Show V_{pp} and T values.

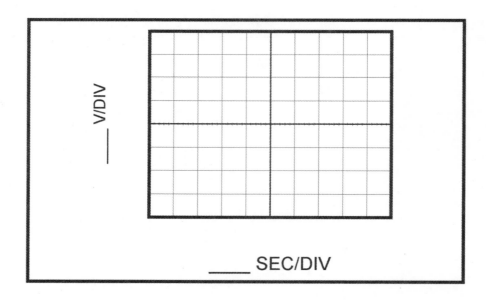

FIGURE 2I-7 Ripple Voltage (AC coupling)

11. Reduce the capacitor from 100 μF to 47 μF and observe the ripple voltage. Did this cause the ripple to increase or decrease? ☐ Increase ☐ Decrease

By approximately how much (half, double)? _____

12. Reduce the load resistor from 2.2 kΩ to 1 kΩ and observe the ripple voltage. Did this cause the ripple to increase or decrease? ☐ Increase ☐ Decrease

By approximately how much (half, double)? _____

13. Change the oscilloscope back to the DC mode, readjusting the sensitivity to observe the total voltage with the 47-μF capacitor and the 1-kΩ load resistor in the circuit.

14. Sketch the output voltage waveform on Figure 21-8 with $R_L = 1$ kΩ and $C = 47$ μF.

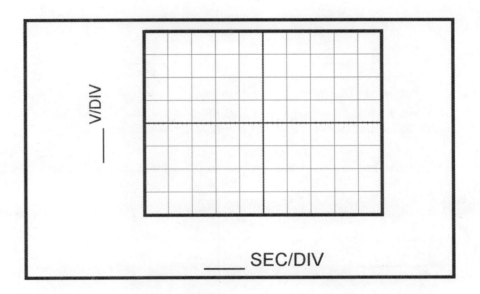

FIGURE 21-8 Filtered Full-Wave Output Voltage (with $R_l = 1$ kΩ and $C = 47$ μF)

15. Using this oscilloscope observation, measure V_{out} peak output voltage, peak-to-peak ripple voltage, and the minimum voltage and record them in Table 21-5.

TABLE 21-5 Filtered Full-Wave Output Voltage with $R_L = 1$ kΩ and $C = 47$ μF

Quantity	Expected Value	Measured Value	% Error
$V_{out\,p}$			
$V_{out\,pp\,ripple}$			
$V_{out\,min}$			
$V_{out\,dc}$			

SAMPLE CALCULATIONS

Show your pre-lab calculations for:

$V_{out\ pp\ ripple} =$

$V_{out\ dc} =$

---

16. Using the DMM, measure the $V_{out\ dc}$ output voltage. Record the measurement in Table 21-5.

17. Draw a dark, solid line through the waveform in Figure 21-8 representing the measured DC value. Does this DC value make sense relative to the peak and minimum values of Figure 21-8?

☐ Yes ☐ No

18. Is the filtered peak output voltage the same as it was in the unfiltered case?   ☐ Yes ☐ No

Observations   ■

1. What is the function of the capacitor in the power supply?

_____

_____

2. How is the ripple voltage affected if either the value of C or R_L is reduced?

_____

_____

Procedure 21-4

Constructing a Power Supply: Zener Regulation

1. Measure and record component values in Table 21-6.

TABLE 21-6 Component Measurements for Circuit of Figure 21-9			
Component	Nominal Value	Measured Value	% Error
R_S	330 Ω		
R_L	1 kΩ		
C	47 μF		

2. Power down the circuit. Connect the circuit of Figure 21-9. Be sure to connect your zener in the reverse-bias direction.

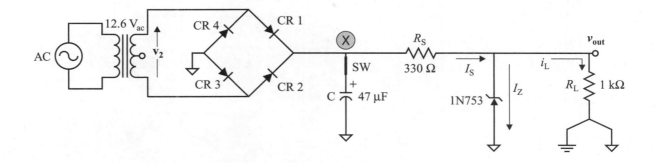

FIGURE 21-9 Zener Voltage-Regulated Power Supply

3. Observe the filtered voltage V_X and sketch it in Figure 21-10. Record its maximum value V_{Xp} and its minimum value $V_{X(min)}$ in Table 21-7 on page 274.
4. Observe the regulated voltage V_{out} and sketch it in Figure 21-10. Clearly label both waveforms.

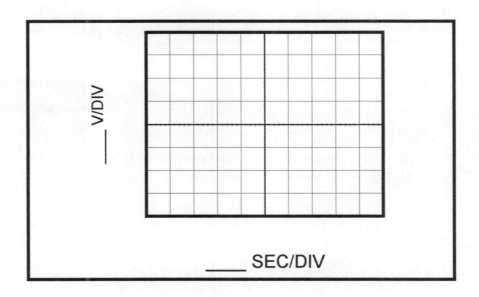

FIGURE 21-10 Zener Regulated Power Supply Output Voltage (DC mode)

TABLE 21-7 Zener-Voltage Regulated Power Supply

Quantity	Expected Value	Measured Value	% Error
$V_{out\ dc}$		(STOP)	
$I_{L\ dc}$			
P_L		**CAREFUL** power rating	
V_{Xp}			
$V_{RS(max)}$			
$I_{RS(max)}$			
V_{Xpp} of the ripple			
$V_{out\ pp}$ of the ripple		(STOP)	
$V_{X\ dc}$			
$V_{RS\ dc}$			
$I_{RS\ dc}$			
P_{RS} use DC values		**CAREFUL** power rating	
$I_{Z(max)}$		**CAREFUL** I_z max rating	
$P_{Z(max)}$		**CAREFUL** power rating	
$V_{X(min)}$			
$I_{RS(min)}$			
$I_{Z(min)}$		Enough to hold regulation?	

(max)	Assume **peak voltage across capacitor** being applied to R_S.
(min)	Assume **minimum voltage across capacitor** being applied to R_S

(STOP)   Instructor sign-off of measured values in Table 21-7 _____

5. Change the oscilloscope to the AC mode and increase the sensitivity to observe any possible ripple. Is there still a ripple in the output?

               ☐ Yes ☐ No

6. Roughly sketch the waveform on Figure 21-11, indicating its measured peak-to-peak value.

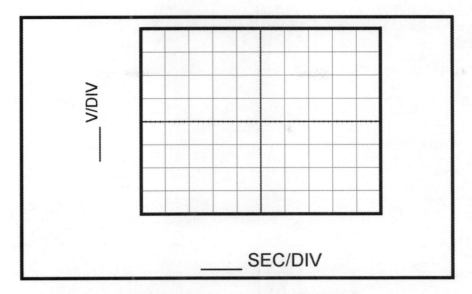

FIGURE 2I-II   Zener Regulated Power Supply Output Voltage (AC mode)

7. Measure and record the remaining values in Table 21-7. Use the appropriate test instrument to measure voltage and current. Calculate power from the measured voltage and current values.

SAMPLE CALCULATIONS

Show your pre-lab calculations for:

$I_{RS\ dc} =$

$V_{out\ dc} =$

$P_L =$

---

8. Based upon your values in Table 21-7, are any of your components exceeding their limits in power or current?

□ Yes □ No

Observations

1. Are any components in danger of failure due to excessive power or current?

_____

_____

2. What should you do if they are?

_____

_____

3. What is the function of the zener diode in the power supply?

_____

_____

4. How is the load ripple voltage affected by the presence of the zener diode in the circuit?

_____

_____

Synthesis

Selecting R_S and Observing Zener Dropout

First, change R_S from 330 Ω to the lowest possible resistance for which the zener diode has not exceeded its power rating. Note that the worst-case zener current occurs when R_L is an open and all current flows through R_S *and* through the zener diode. R_S must limit the current for this case.

Then change R_L to 180 Ω and change C from 47 µF to 10 µF as shown in Figure 21-12. Measure and record component values in Table 21-8.

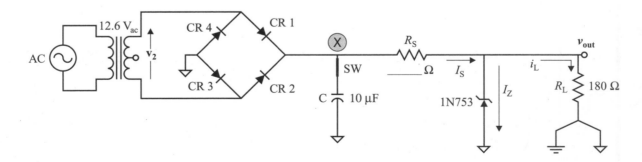

FIGURE 21-12 Zener Voltage-Regulated Power Supply (Zener dropout)

TABLE 21-8   Component Measurements for Circuit of Figure 21-12

Quantity	Nominal Value	Measured Value	% Error
R_S			
R_L	180 Ω		
C	10 μF		

Calculate the expected values to determine if there might be any component-rating problems. Be sure to use appropriate components to ensure safe operation.

SAMPLE CALCULATIONS

Show your pre-lab calculations for:

$I_{L\ dc} =$

$V_{RS(max)} =$

$V_{Xpp} =$

In your calculations do you see any potential zener dropout problem? Is there a problem with your expected peak-to-peak ripple voltage calculation? If so, what is it?

_____

_____

Observe the filtered voltage V_X and sketch it in Figure 21-13. Record its maximum value V_{Xp} and its minimum value $V_{X(min)}$ in Table 21-9. Observe the regulated voltage V_{out} and sketch it in Figure 21-13. Clearly label both waveforms.

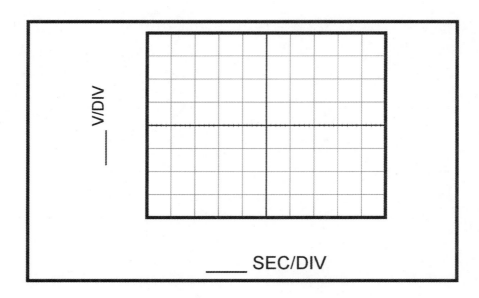

V/DIV

_____ SEC/DIV

FIGURE 21-13 Zener Regulated Power Supply Output Voltage (Zener dropout)

TABLE 21-9 Zener Voltage-Regulated Power Supply (Zener dropout)

Quantity	Expected Value	Measured Value
$V_{out\ dc}$		
$I_{L\ dc}$		
P_L		
V_{Xp}		
$V_{RS(max)}$		
$I_{RS(max)}$		
V_{Xpp} of the ripple		

Observations

1. What is the purpose of the series R_S resistor?

   _____

   _____

2. What are some situations that might cause the zener to drop out of regulation?

   _____

   _____

Computer Activity

Simulate the circuit of Figure of 21-8. Use the sine-wave voltage with 17.8 volts peak at 60 Hz in place of the transformer in the actual circuit. Compare the results of the simulation with the actual results obtained. How are the two results similar? How are they different?

_____

_____

_____

_____

_____

_____

Three-Terminal IC Regulator

Name: _____ Date: _____

Lab Section: _____ _____ Lab Instructor: _____
              day   time

Text Reference

DC/AC Circuits and Electronics: Principles and Applications
Chapter 18: Power Supply Applications

Materials Required

Triple power supply (2 @ 0–20 volts DC; 1 @ 5 volts DC)

1 each 0.1-μF, 470-μF capacitor

5 1N4001 rectifier diodes

1 LM 340-5 three-terminal IC regulator

1 120 to 12.6-V transformer with center-tapped secondary

1 Heat sink

Introduction

In this exercise, you will:

- Examine an IC voltage regulator under varied load conditions
- Examine the aspect of current limiting of the IC regulator
- Examine the effect of thermal shutdown of the IC
- Examine the effect of using a heat sink on the regulator IC
- Examine how to design an IC-regulated power supply.

Pre-Lab Activity Checklist ■

☐ Find the expected values for Table 22-1 and include sample calculations.
☐ Build the circuit of Figure 22-2.

Performance Checklist ■

☐ Pre-lab completed?　　　　　　　　　🛑 Instructor sign-off _____

☐ Table 22-1: Demonstrate measured $V_{out\ dc}$ and $V_{reg\ (in-out)\ dc}$.

☐ Table 22-2: Demonstrate measured $V_{out\ dc}$ and $V_{reg\ (in-out)\ dc}$.

☐ Procedure 22-3: Demonstrate regulator operation under load.

Procedure 22-1

Voltage Regulation Using an IC Fixed Regulator

> ⚠ **WARNING!** This regulator can attain a maximum temperature of 150°C. Do not touch it while it is on or immediately after the circuit is powered off.

1. In the circuit of Figure 22-1, vary R_{load} to set the output current to 100 mA. Use the oscilloscope and the DMM to measure or observe the voltages.

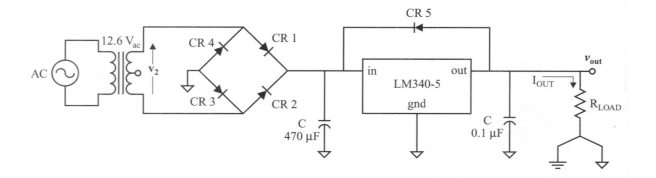

FIGURE 22-1　Power Supply Using a Fixed IC Voltage Regulator

2. Record the quantities in column 3 of Table 22-1 for an output current of 100 mA. Calculate the value for P_D and T_J from the measured values. Perform the calculations for P_D, T_J, and T_{case} in the space provided after Table 22-1 in the Sample Calculations box.

TABLE 22-1 Power Supply Measurements (*calculated from measured values)

	Test Instrument Used	Expected Value	Measured Value	Measured Value	Measured Value	Measured Value
$I_{out\ dc}$		100 mA	100 mA	200 mA	300 mA	400 mA
$V_{out\ dc}$			(STOP)			
$V_{out\ pp\ ripple}$						
$V_{out\ rms}$						
$V_{cap\ ac\ peak}$						
$V_{cap\ min}$						
$V_{cap\ pp\ ripple}$						
$V_{470\ \mu F\ dc}$						
$V_{0.1\ \mu F\ dc}$						
$V_{reg\ (in\text{-}out)\ dc}$			(STOP)			
$P_{D\ (regulator)}$						
$T_A\ °C$						
$T_J\ *\ °C$						
$T_{case}\ °C$						

(STOP) Instructor sign-off of measured values in Table 22-1 _____

SAMPLE CALCULATIONS

These calculations are based on measured values for the 100-mA load current condition.

Regulator power dissipation $P_D =$

Regulator junction temperature $T_J =$

Case temperature $T_{case} =$

---

3. Increase the output current in 100 mA steps and complete the measurements in Table 22-1 for each load current *until* thermal shutdown occurs. Perform calculations for P_D, T_J, and T_{case} for each output current before raising the output current to the next level. Watch for indicators of thermal shutdown, which are a slight drop in DC output voltage and a sudden drop in output current.

   ◢ **Note:** Once thermal shutdown occurs, do not increase I_{out}. Leave the remaining columns in Table 22-1 blank.

4. To confirm that the regulator has gone into thermal shutdown, fan or blow on the regulator to cool it. Observe the output voltage with one DMM and the output current with a second DMM. Simultaneously observe the output waveform with an oscilloscope. As the regulator cools, the output current should return to its previous level.

5. While the regulator is in thermal shutdown, grasp the regulator with pliers and observe the effect on the output voltage.

Observations ■

1. Describe the effect of cooling the regulator on the output.

   _____

   _____

2. Why would one use an IC regulator rather than a zener diode?

   _____

   _____

3. Under the highest load current condition prior to thermal shutdown, determine the ripple rejection.

   _____

4. What is the ripple rejection in dB?

   _____

5. Describe the effect of grasping the regulator with pliers, thereby adding a heat sink on the output.

_____

_____

_____

_____

Procedure 22-2

Heat Sink Effect

1. Place a heat sink on the regulator. **BE CAREFUL:** The regulator is very hot; use pliers to hold it while attaching the heat sink.
2. Remeasure the power supply quantities at which thermal shutdown previously occurred. Record the values in the second column of Table 22-2.
3. Increase load current by 100 mA and record the values in the third column of Table 22-2.

TABLE 22-2 Power Supply Measurements with Heat Sink (*calculated from measured values)		
	Measured Values	**Measured Values**
$I_{out\ dc}$		
$V_{out\ dc}$		
$V_{out\ pp\ ripple}$		
$V_{out\ rms}$		
$V_{cap\ ac\ peak}$		
$V_{cap\ min}$		
$V_{cap\ pp\ ripple}$		
$V_{470\ \mu F\ dc}$		
$V_{0.1\ \mu F\ dc}$		
$V_{reg\ (in-out)\ dc}$		
$P_{D\ (regulator)}$		
$T_A\ °C$		
$T_J*\ °C$		
$T_{case}\ °C$		

🛑 Instructor sign-off of measured values in Table 22-2 _____

Observations

1. What was the increase in the available load current with the heat sink? _____

2. What was the increase in the amount of power dissipation of the regulator with the heat sink added?

    _____

3. Based upon your results, predict the maximum current that can be delivered to the load with a heat sink before thermal shutdown.

    _____

    _____

4. Would current limiting come into play before thermal shutdown? ☐ Yes ☐ No

Synthesis
Design a Regulated Voltage Source

Use an LM340-5 voltage regulator to design, build, and test a filtered, voltage-regulated power supply adjustable from 5 V to 9 V at 0.3 A. Calculate the quantities requested below and fill-in each of the blanks below. Calculate the values necessary as directed and then build and test your circuit design.

Transformer and Capacitor Specifications

1. What is the specific part number of your LM340-5? _____

2. What package style is your regulator? _____

3. What is the regulator voltage dropout specification? _____

4. What is the regulator minimum input voltage allowed? _____

5. What is the regulator maximum input voltage specification? _____

6. Based upon the above information, sketch the worst case ripple voltage permitted to the input of this voltage regulator.

7. Using this information, determine which transformer (12.6 V or 25.2 V), rectifier circuit, and capacitor you will use. Technically justify your choices.

Rectifier circuit _____

Transformer _____

Capacitor _____

Adjustment Resistor Selection

1. Design the voltage regulator resistor circuit to produce an adjustable regulated voltage from 5 V to 9 V. Draw it below.

2. What amount of power will the adjustment resistors need to dissipate? _____

Re-Evaluate the Input Voltage to the Voltage Regulator

1. Now, based upon your load current, adjustment resistor current, capacitor value, and transformer data curves, sketch the expected regulator input waveform.

2. Are all input regulator requirements still met? ☐ Yes ☐ No

Heat Dissipation

1. Is the regulator working the hardest when V_{out} is 5 V or 9 V? ☐ 5 V ☐ 9 V
2. Given the above design, what is the maximum power the IC will be required to dissipate?

   _____

3. Assuming no heat sink, what is the manufacturer's specification for θ_{JA}? _____

4. What is a reasonably expected T_A? _____

5. What is the calculated IC junction temperature T_J? _____

6. What is the T_{Jmax} specification? _____

7. Does this IC regulator require a heat sink? ☐ Yes ☐ No

Draw Your Designed Circuit Schematic.

Construct Your Circuit

Be sure your circuit meets the original specifications. Consider the following questions before applying power to your circuit:

- What are you doing about heat sinking?
- What power will the load have to dissipate?
- What will you use as a load?

Test your design and determine if it meets specifications, then demonstrate the working circuit to your instructor.

🛑 Instructor sign-off _____

Inverting and Noninverting Operational Amplifiers

Name: _____ Date: _____

Lab Section: _____ _____ Lab Instructor: _____
                   day   time

Text Reference

DC/AC Circuits and Electronics: Principles and Applications
Chapter 12: Waveforms
Chapter 19: Dependent Sources

Materials Required

Triple power supply (2 @ 0–20 volts DC; 1 @ 5 volts DC)

1   each 1-kΩ, 10-kΩ, 22-kΩ, 1/4-watt resistor

2   0.1-μF capacitors

1   uA741 operational amplifier

Introduction

In this exercise, you will:

- Examine a noninverting voltage amplifier
- Examine an inverting voltage amplifier
- Examine a summing amplifier
- Examine a two-stage amplifier.

Pre-Lab Activity Checklist

☐ Find the expected values for Table 23-2 and include sample calculations.

☐ Find the expected values for Table 23-4 and include sample calculations.

☐ Find the expected values for Table 23-6 and include sample calculations.

☐ Build the circuit of Figure 23-1. **Note:** The 50-Ω source resistor is internal to the function generator. Do not add an external resistor.

Performance Checklist

☐ Pre-lab completed?   🛑 Instructor sign-off _____

☐ Table 23-2: Demonstrate measured V_{in} and V_{out} of the AC signal.

☐ Table 23-4: Demonstrate measured V_{in} and V_{out} of the AC signal.

☐ Table 23-6 and Figure 23-4: Demonstrate measured V_1, V_{out} of DC signal, and the V_{out} waveform.

Tutorial

Operational Amplifier Board Layout Guidelines

1. Do not use busses for connecting power to the IC. Run power directly to the V+ and V− pin of the IC with the shortest possible wire. Busses provide an antenna for receiving and spreading electromagnetic interference. The 0.1- to 0.5-Ω resistance and 75-ηH/inch inductance will devastate the circuit's performance.

2. There should be only one ground point in the circuit and that is at the power supply common. Connect all common runs to that one point. All instrument commons must be tied directly to the single common point. A 5-mV difference in a ground line can be comparable to the input signal.

3. You may use a ground bus or ground plane, but only for those signals that carry less than 50-mA of current. Otherwise, run the return directly to the single point ground at the power supply. This is particularly true for loads; they must be returned to the power supply common in their own lead.

4. Decoupling capacitors (those capacitors connected from the power supply to common) should be connected directly to the power pin of the IC and directly to ground (ground plane if you have one). Trim the leads of the capacitors as short as practical to minimize lead inductance. Decoupling the power bus does little good. Decouple the IC power pin.

5. Keep lead length into the op amp's inputs as short as possible. Electromagnetic noise fields intersecting these leads generate noise currents that flow into the large input impedance of the op amp and cause significant problems.

Use of the above guidelines will ensure proper circuit performance for all op-amp circuits.

Procedure 23-1

Noninverting Voltage Amplifier

1. Measure and record the component values of Figure 23-1 in Table 23-1.

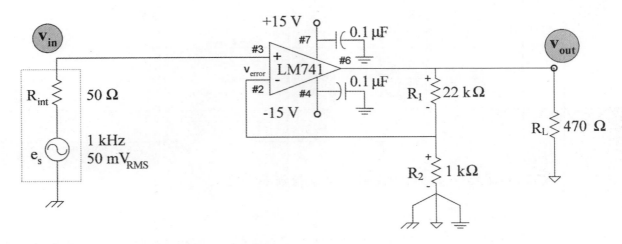

FIGURE 23-I Noninverting Voltage Amplifier (*pre-lab*)

TABLE 23-I Component Values for the Noninverting Voltage Amplifier

Component	Nominal Value	Measured Value
R_1	22 kΩ	
R_2	1 kΩ	
R_L	470 Ω	

2. Using the DMM, set the unloaded function generator AC signal, e_s, to 50 mV and set the DC voltage, E_S, to 0 V. Set the frequency to 1 kHz with either a frequency counter or an oscilloscope. Record these measured voltages in Table 23-2 on page 294.

3. Connect the function generator to the circuit, and connect the oscilloscope to observe the input signal voltage on CH1 (the upper half of the oscilloscope display) and the output signal voltage on CH2 (the lower half of the oscilloscope display).

4. Measure and record the quantities of Table 23-2 using the DMM. Use the oscilloscope to determine the phase relationship of v_{out} with respect to v_{in}.

TABLE 23-2 Characteristics of the Noninverting Voltage Amplifier

Quantity	Expected Values	Measured Values	
E_s unloaded			
e_s unloaded			
V_{in}			
$v_{in\ AC}$			STOP
V_{out}			
v_{R2}			
$v_{out\ AC}$			STOP
v_{out} vs. v_{in} phase			

STOP Instructor sign-off of measured values in Table 23-2 _____

SAMPLE CALCULATIONS
_____

Show your pre-lab calculations for:

$V_{out\ ac} =$

Observations ■

1. Calculate the AC signal expected voltage gain based on the component values in the circuit.

Expected A_v _____

2. Calculate the measured voltage gain and compare this value with the expected value above.

Measured A_v _____

3. Do these values agree?　　　　　　　　　　　　□ Yes　□ No
4. Are the input and output in phase or out of phase?　　□ In phase　□ Out of phase

Procedure 23-2

Inverting Voltage Amplifier

1. Following the op-amp board layout guidelines, construct the circuit of Figure 23-2.

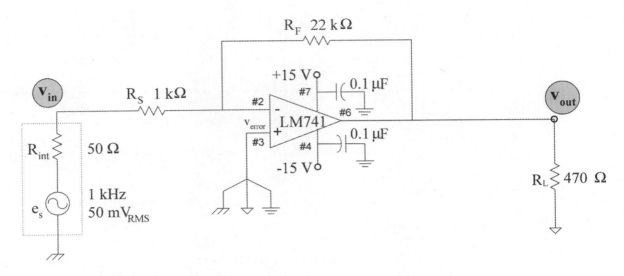

FIGURE 23-2 Inverting Voltage Amplifier

2. Measure and record component values in Table 23-3.

TABLE 23-3 Component Values of the Inverting Voltage Amplifier

Component	Nominal Value	Measured Value
R_S	22 kΩ	
R_F	1 kΩ	
R_L	470 Ω	

3. Using the DMM, set up the unloaded function generator AC signal, e_s, to 50 mV. Set the DC voltage, E_S, to 0 V. Set the frequency to 1 kHz with either a frequency counter or oscilloscope. Record these measured voltages in Table 23-4 on page 296.

4. Connect the function generator to the circuit, and connect the oscilloscope to observe the input signal voltage on CH1 (the upper half of the oscilloscope display) and the output signal voltage on CH2 (the lower half of the oscilloscope display).

5. Measure and record the quantities of Table 23-4 using the DMM. Use the oscilloscope to determine the phase relationship of v_{out} with respect to v_{in}.

TABLE 23-4 Characteristics of the Inverting Voltage Amplifier

Quantity	Expected Values	Measured Values	
E_S			
e_s			
V_{in}			
v_{in}			(STOP)
$V_{pin\ \#2}$			
V_{out}			
$v_{pin\ \#2}$			
v_{out}			(STOP)
v_{out} vs. v_{in} phase			

(STOP) Instructor sign-off of measured values in Table 23-4 _____

6. Save this circuit; it will be modified for the next procedure.

SAMPLE CALCULATIONS

Show your pre-lab calculations for:

$V_{out\ ac} =$

Observations ◾

1. Calculate the expected voltage gain based on the component values in the circuit.

Expected A_v _____

2. Calculate the measured voltage gain and compare this value with the expected value.

Measured A_v _____

3. Do these values agree?   ☐ Yes ☐ No
4. Are the input and output in phase or out of phase?   ☐ In phase ☐ Out of phase

Procedure 23-3
Summing Voltage Amplifier

1. Measure resistor R_1 to be used in the circuit of Figure 23-3 and record its value in Table 23-5.

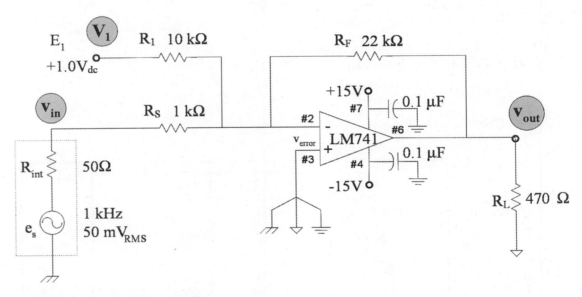

FIGURE 23-3 Summing Voltage Amplifier

TABLE 23-5 Component Value for the Summing Voltage Amplifier		
Component	**Nominal Value**	**Measured Value**
R_1	10 kΩ	

2. Set up the unloaded DC power supply, E_1, to +1.0 V and record its measured value in Table 23-6 on page 298.
3. Add the DC input circuit (E_1 and R_1) to the circuit of completed Figure 23-3.

TABLE 23-6 Characteristics of the Summing Voltage Amplifier

Quantity	Unit	Expected Value	Measured Value
E_1	V_{dc}		
V_1	V_{dc}		STOP
$V_{pin\ \#2}$	V_{dc}		
V_{out}	V_{dc}		STOP
e_s	V_{rm}		
v_{in}	V_{rm}		
v_{pin}	V_{rm}		
v_{out}	V_{rm}		

4. Measure and record the remaining quantities of Table 23-6 using the DMM. Use the oscilloscope to determine the phase relationship of v_{out} with respect to v_{in}.

5. Sketch the total v_{out} waveform on Figure 23-4.

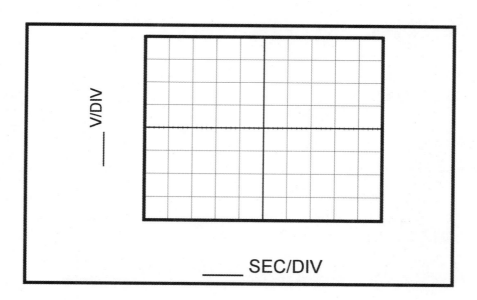

FIGURE 23-4 Total v_{out} Waveform of Summing Voltage Amplifier

STOP Instructor sign-off of measured values in Table 23-6 and Figure 23-4 _____

SAMPLE CALCULATIONS

Show your pre-lab calculations for:

$V_{\text{out dc}} =$

Observations

1. Based upon your measured values, did this amplifier perform as a summing amplifier?

    ☐ Yes   ☐ No

2. Calculate the DC signal expected voltage gain from V_1 to V_{out} based on the component values in the circuit.

    Expected A_{v1} _____

3. Calculate the measured DC signal voltage gain from V_1 to V_{out} and compare this value with the expected value above.

    Measured A_{v1} _____

4. Do these values agree?

    ☐ Yes   ☐ No

5. Calculate the AC signal expected voltage gain from v_{in} to v_{out} based on the component values in the circuit.

    Expected A_{vin} _____

6. Calculate the measured voltage gain and compare this value with the expected value above.

    Measured A_{vin} _____

7. Do these values agree?

    ☐ Yes   ☐ No

Synthesis

Design of a Two-Stage Amplifier

Design a two-stage amplifier that has a gain of 21.3 from the input of the first stage to the output of the second stage. The input signal to the first stage and the output signal of the second stage must be in phase (noninverting) relative to each other.

Neatly draw the schematic for your circuit in the space for Figure 23-5. Select values for the input and feedback resistors from those available in the parts required for this exercise. Measure those components and enter their values in Table 23-7.

FIGURE 23-5   Schematic of Two-Stage Noninverting Amplifier; $A_v = 21.3$

Build your circuit. Measure the output signal amplitude and record the measured gain in Table 23-7. Then measure the individual stages gains. Observe the phase relationships between the various stages of the amplifier.

TABLE 23-7   Two-Stage Amplifier Results

First Stage			Second Stage			Total
R_i	R_f	A_V	R_i	R_f	A_V	A_V

Observations

1. What was the percent difference between your gain and the desired gain of 21.3?

% difference = _____

2. Was the output signal in phase with the input? ☐ Yes ☐ No

Why or why not? _____

_____

Computer Activity

1. Simulate the circuit of Figure of 23-1. Use the sine-wave voltage source as the signal source with 50-mV amplitude and a frequency of 1 kHz. Compare the results of the simulation with the actual results obtained.

How are the two results similar? _____

_____

How are they different? _____

_____

2. Simulate the circuit of Figure of 23-2. Compare the results of the simulation with the actual results obtained.

How are the two results similar? _____

_____

How are they different? _____

_____

Amplifier Impedance and Modeling

Name: _____   Date: _____

Lab Section: _____ _____   Lab Instructor: _____
                        day   time

Text Reference ◼

DC/AC Circuits and Electronics: Principles and Applications
Chapter 19: Dependent Sources

Materials Required ◼

Triple power supply (2 @ 0–20 volts DC; 1 @ 5 volts DC)
1 each 1-kΩ, 1.2-kΩ, 2.2-kΩ, 3.3-kΩ, 10-kΩ, 12-kΩ, 100-kΩ, 1/4-W resistor
3 0.1-μF capacitors
3 10-μF capacitors, minimum 35-W Vdc
1 uA741 operational amplifier
1 2N3904 NPN transistor

Introduction ◼

In this exercise, you will:

- Examine a function generator's internal resistance
- Examine the input impedance, output impedance, and voltage transfer function of a common emitter amplifier
- Examine the input impedance, output impedance, and voltage transfer function of an inverting voltage amplifier
- Examine the voltage transfer function of a multi-stage amplifier.

Pre-Lab Activity Checklist ■

□ Devise and write a test procedure to experimentally measure the internal output impedance of the function generator including a schematic of your test circuit and a data table to record your measured values.

□ Find the expected values in Tables 24-2, 24-3, 24-4, 24-5, 24-7, 24-8, and 24-9. Include sample calculations.

□ Build the circuit of Figure 24-1 on the left side of your protoboard.

□ Build the circuit of Figure 24-3 on the right side of your protoboard.

□ Draw the AC model of the circuit of Figure 24-4 in Figure 24-5.

Performance Checklist ■

□ Pre-lab completed?   🛑 Instructor sign-off _____

□ Table 24-3: Demonstrate measured V_b, V_e, V_c.

□ Table 24-5: Demonstrate measured Z_{in}, Z_{out} at C.

□ Table 24-7: Demonstrate measured V_{in}, V_{out}.

□ Table 24-9: Demonstrate measured A_{V1}, A_{V2}.

Procedure 24-1

Generator Internal Resistance

1. Perform your experimental procedure for the function generator output and record results in your table(s).

Observation ■

1. What is the measured internal resistance of your function generator? _____

Procedure 24-2

Common Emitter Amplifier: Transfer Function A_v

1. Measure and record the component values of Figure 24-1 in Table 24-1.

2. Adjust the loaded output of the DC power supply to 15 V.

3. Adjust the unloaded function generator AC signal. Use the DMM to read the output voltage while making this adjustment. Adjust the frequency with an oscilloscope, frequency counter, or DMM.

4. Check the orientation of both electrolytic capacitors.

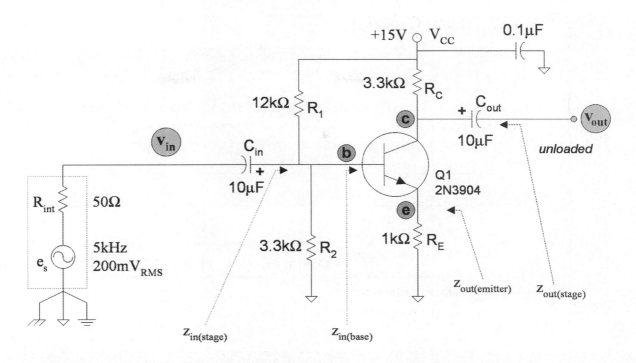

FIGURE 24-I   Schematic of a Common Emitter Amplifier (*pre-lab*)

TABLE 24-I   Component Values Used in Figure 24-I

Component	Nominal Value	Measured Value
C_{in}	10 μF	
C_{out}	10 μF	
$C_{decouple}$	0.1 μF	
R_1	12 kΩ	
R_2	3.3 kΩ	
R_C	3.3 kΩ	
R_E	1.0 kΩ	

5. Connect the signal generator to the circuit. Measure and record the DC voltages of Table 24-2. Do not continue if the measured quiescent DC voltages do not meet expectations. The circuit will not work properly if the DC voltages are not correct.

TABLE 24-2 DC Voltages for Figure 24-1

Quantity	Expected V_{dc}	Measured V_{dc}
V_{CC}		
V_B		
V_E		
V_C		

SAMPLE CALCULATIONS

Show your pre-lab calculations for:

$V_E =$

---

6. Measure and record the AC voltages of Table 24-3.

TABLE 24-3 AC RMS Voltages for Figure 24-1

Quantity	Expected V_{rms}	Measured V_{rms}	
v_{cc}			
v_{in}			
v_b			**STOP**
v_e			**STOP**
v_c			**STOP**
v_{out}			

STOP Instructor sign-off of measured values in Table 24-3 _____

SAMPLE CALCULATIONS

Show your pre-lab calculations for:

$V_c =$

7. Measure both the input signal and the output signal simultaneously, using the oscilloscope. You should always observe the AC signals to ensure that there is no unexpected distortion.

8. Calculate and record the quantities of Table 24-4 based upon your measured values.

TABLE 24-4 AC Voltage Gains for Figure 24-1

Quantity	Expected Value	Measured Value
A_v **b** to **c**		
A_v **b** to **e**		

SAMPLE CALCULATIONS

Show your pre-lab calculations for:

$A_{V\,b\,to\,c} =$

Observations ■

1. What is the collector voltage phase with respect to the input?   ☐ In phase ☐ Out of phase
2. What is the emitter voltage phase with respect to the input?   ☐ In phase ☐ Out of phase
3. What is the collector voltage phase with respect to the emitter?   ☐ In phase ☐ Out of phase

Procedure 24-3
Common Emitter Amplifier:
Input and Output Impedances

1. Use the appropriate procedure and test resistor, R_{test}, to find the input impedance to the amplifier stage. Record the values in Table 24-5.

TABLE 24-5 AC Input and Output Impedances for Circuit of Figure 24-1

Quantity	Expected Value	Measured Value
z_{in}		STOP
z_{out} at C		STOP
z_{out} at E		

STOP Instructor sign-off of measured values in Table 24-5 _____

SAMPLE CALCULATIONS

Show your pre-lab calculations for:

Z_{out} at $C =$

---

2. Use the appropriate procedure and test resistor, R_{test}, to find the output impedance at the collector. Record the values in Table 24-5.

3. Use the appropriate procedure and test resistor R_{test}, to find the output impedance at the emitter. Record the values in Table 24-5. The circuit should be configured as shown in Figure 24-2. You must use an AC coupling capacitor from your emitter to your test resistor, R_{test}, to prevent DC loading. Use a 470-μF electrolytic capacitor as a coupling capacitor to your test resistor.

Hint: To prevent loading, use the unmatched method with $p = 0.80$ e_{oc}. The matched method tends to produce excessive circuit loading and mismeasurement in this case.

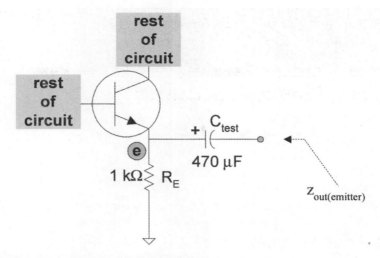

FIGURE 24-2 Test Point for the Emitter Output Impedance

Observation

1. What one characteristic of the transistor is responsible for the input impedance being large compared to the output impedance?

Procedure 24-4
Inverting Voltage Amplifier: Transfer Function

1. Measure the component values for Figure 24-3 and record them in Table 24-6.

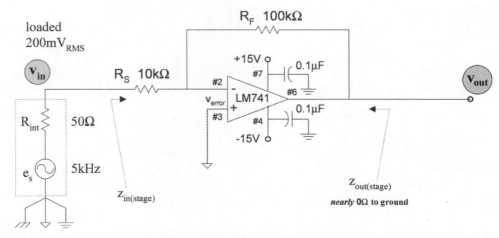

FIGURE 24-3 Schematic of an Inverting Voltage Amplifier (*pre-lab*)

TABLE 24-6 Component Values for the Circuit of Figure 24-3

Component	Nominal Value	Measured Value
R_S	10 kΩ	
R_F	100 kΩ	

2. Adjust the unloaded DC power supply voltages to +/–15 V.

3. Adjust the loaded function generator AC signal. Accurately set up the AC signal voltage using the DMM. Adjust the frequency using an oscilloscope, frequency counter, or DMM.

   ❑ **Note:** Maintain the loaded v_{in} at 200 mV_{rms} throughout this procedure.

4. Measure both the input signal and the output signal simultaneously, using the oscilloscope. Record the phase relationship of the input and output signals in Table 24-7.

5. Measure and record the quantities of Table 24-7 using the DMM. The oscilloscope was already used to determine the phase relationship of v_{out} with respect to v_{in}.

TABLE 24-7 Measurements for the Circuit of Figure 24-3

Quantity	Expected Value	Measured Value
E_S	0 V$_{DC}$	
V_{in}	0 V$_{DC}$	
$V_{pin \#2}$	0 V$_{DC}$	
V_{out}	0 V$_{DC}$	
e_s		
v_{in}		STOP
$v_{pin \#2}$		
v_{out}		STOP
A_v		
Phase of v_n/v_{out}		

STOP Instructor sign-off of measured values in Table 24-7 _____

SAMPLE CALCULATIONS

Show your pre-lab calculations for:

$V_{out\,(rms)} =$

Observation

1. What is the calculated current through R_s? Show your equation.

Procedure 24-5
Inverting Voltage Amplifier: Input Impedance, Output Impedance, and Max Load Current

1. Use the appropriate procedure and test resistor, R_{test}, to find the amplifier input impedance to the amplifier stage. Record the values in Table 24-8.

TABLE 24-8 AC Input and Output Impedances for the Circuit of Figure 24-3

Quantity	Expected Value	Measured Value
z_{in} of circuit		
z_{out} of circuit	~0 Ω	
max I_{out} of op amp	*	

* Device specification

SAMPLE CALCULATIONS

Show your pre-lab calculations for:

$Z_{in}=$

---

2. Use the appropriate procedure and test resistor to find the amplifier output impedance. Why can you not effectively perform this procedure in lab?

_____

3. Determine the maximum output current by placing a short from the V_{out} terminal to ground. Do this by placing the DMM in the ammeter mode and placing it across the output terminal to ground. Remember, the ammeter is equivalent to a short.

Observation ■

1. Was the input impedance relatively large or small? □ Large □ Small
2. Verify that the DC values of each stage are still the same. If not, correct the problem before you continue.
3. Measure and record quantities in Table 24-9.
4. Measure signals v_{in}, $v_{out\ 1}$, and v_{out} and sketch the model of the circuit in Figure 24-5.

Synthesis
Cascaded Amplifier

1. Construct the cascade amplifier of Figure 24-4 by connecting the bipolar junction transistor (BJT) amplifier circuit of Figure 24-1 to the operational amplifier circuit of Figure 24-3. Connect the load resistor and readjust the voltage generator as specified to complete the circuit setup.

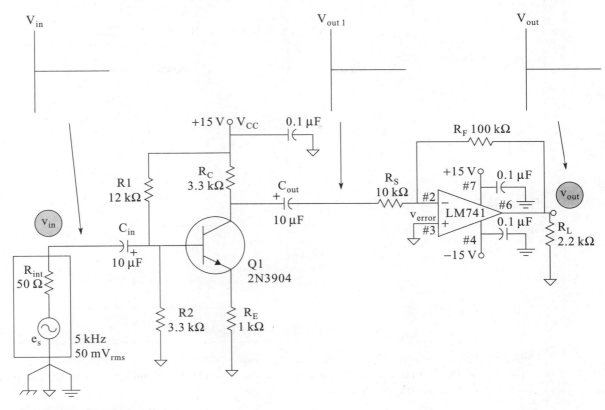

FIGURE 24-4 Cascade Amplifier

TABLE 24-9 Cascade Amplifier Measurements

Quantity		Unit	Expected Value	Measured Value
R_L		Ω		
v_{in}		V_{rms}		
$v_{out\,1}$		V_{rms}		
v_{out}		V_{rms}		
A_{v1}	stage 1			**STOP**
A_{v2}	stage 2			**STOP**
A_v amplifier	v_{out}/v_{in}			

STOP Instructor sign-off of measured values in Table 24-9 _____

FIGURE 24-5 VCVS Model of the Cascade Amplifier (*pre-lab*)

SAMPLE CALCULATIONS

Show your pre-lab calculations for:

$A_{V1} =$

Observation ■

1. What is the output voltage phase with respect to the input?   ☐ In phase ☐ Out of phase

DC Biasing of Basic Amplifiers

Name: _____   Date: _____

Lab Section: _____ _____   Lab Instructor: _____
                day   time

Text Reference

DC/AC Circuits and Electronics: Principles and Applications
Chapter 19: Dependent Sources

Materials Required

Triple power supply (2 @ 0–20 volts DC; 1 @ 5 volts DC)
1   each 1-kΩ, 39-kΩ, and various other standard values of 1/4-W resistor
1   8.2-Ω, 5-W resistor
1   8-Ω speaker
2   each 0.1-μF, 47-μF capacitors
1   each 2N3904 and 2N3055 NPN transistors
1   uA741 operational amplifier

Introduction

In this exercise, you will:

- Examine the effects on an amplifier of biasing an AC input signal with DC offset
- Examine emitter-supply biasing of an amplifier
- Examine voltage-divider biasing of an amplifier
- Examine the effects of biasing an op amp with a split supply system
- Examine a design to allow an op amp to operate with a single supply.

Pre-Lab Activity Checklist

☐ Select a value for R_2 in the circuit of Figure 25-7 to set the base voltage at 2.5 V_{dc}.
☐ Design an external bias circuit for the circuit of Figure 25-13 and draw your circuit in Figure 25-15.
☐ Find the expected values in Tables 25-1 through 25-4 and Table 25-7 and include sample calculations.
☐ Build the circuit of Figure 25-1.

Performance Checklist

☐ Pre-lab completed?   🛑 Instructor sign-off _____
☐ Table 25-1: Demonstrate measured $V_{E1}, V_{C1}, V_{E2}, V_{C2}$.
☐ Table 25-3: Demonstrate measured $V_{C1}, V_{C2}, V_{E1}, V_{E2}$.
☐ Demonstrate the waveforms in Figure 25-6.
☐ Demonstrate the waveforms in Figure 25-8.
☐ Demonstrate the waveforms in Figure 25-17.

Pre-Lab

Procedure 25-1

BJT Amplifier with Source Biasing

1. Adjust the unloaded DC power supply with the indicated collector supply voltage (V_{CC}) for the circuit of Figure 25-1. Accurately adjust the supply using the DMM.

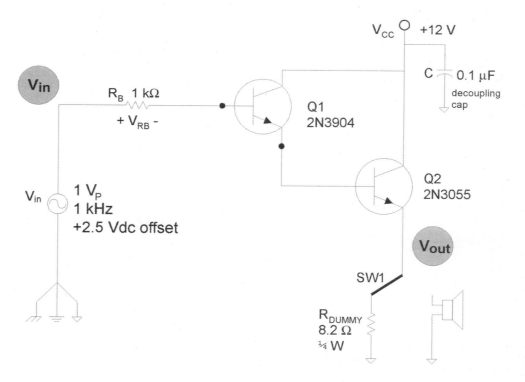

FIGURE 25-1 Amplifier with DC Bias Provided by the Function Generator

2. With its DC offset set to zero, adjust the unloaded function-generator amplitude and period using CH1 of the oscilloscope with DC coupling. Use DC coupling to observe the DC-biasing effects.

3. Adjust the DC offset using the DMM while using the oscilloscope to observe the effect on the signal.

4. Construct the circuit of Figure 25-1 with sources powered off and SW1 positioned to connect to the dummy load resistor. Verify your R and C component values to ensure that you are using the proper components.

5. Connect the DC collector supply voltage and then connect the function generator to the circuit.

   ⊐ **Note:** Proper DC operating voltages should be applied to an active device before a signal is applied. For the same reason, DC operating voltages should be removed only after the signal has been removed from the active device. While this is not critical to all active devices, it is a good practice to observe.

6. Measure and record the circuit DC voltages in Table 25-1. Do not continue if the measured quiescent DC voltages do not meet expectations. The circuit will not work properly if the DC voltages are not correct.

TABLE 25-1 DC Voltages for the Circuit of Figure 25-1

Quantity	Expected V_{dc}	Measured V_{dc}
V_{dc} of input		
V_{CC}		
V_{C1}		(STOP)
V_{C2}		(STOP)
V_{B1}		
V_{E1}		(STOP)
V_{B2}		
V_{E2}		(STOP)

(STOP) Instructor sign-off of measured values in Table 25-1 _____

7. Leave CH1 connected to the input signal and connect CH2 to the emitter of Q_2. Sketch the total output voltage waveform (including the DC offset) on Figure 25-2.

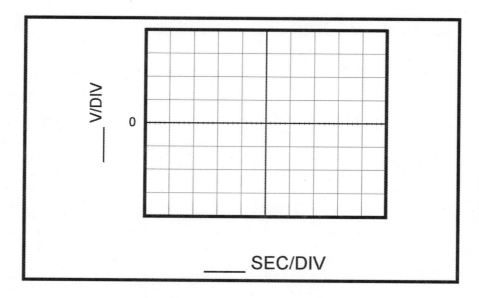

FIGURE 25-2   Amplifier Audio Output Signal with Resistive Load

---

SAMPLE CALCULATIONS

Show your pre-lab calculations for:

⊐ **Note:** For these calculations assume that the beta of Q_1 is 200 and the beta of Q_2 is 30.

$V_E =$

$V_C =$

Function Generator DC Offset Voltage _____

8. Once you observe the proper signal at the output, measure the RMS voltage of the AC output voltage and the RMS voltage of the total output voltage (DC + AC). Record these values in Table 25-2.

TABLE 25-2 RMS AC Only and Total Signal Output Voltages

Quantity	Expected V_{rms}	Measured V_{rms}
AC of V_{out}		
Total of V_{out}		

SAMPLE CALCULATIONS

Show your pre-lab calculations for:

$V_{out\ ac} =$

---

9. Turn off the signal generator and then turn off the DC supply. Using SW1, connect the speaker in place of the dummy load. Turn on the DC power supply and then turn on the signal generator. The signal should be the same.

10. With the DMM DC voltmeter connected to the input signal and CH2 of the oscilloscope still connected to the emitter of Q_2, decrease the function generator DC offset until clipping becomes obvious. Sketch the waveform on Figure 25-3 and record its DC offset voltage below Figure 25-3.

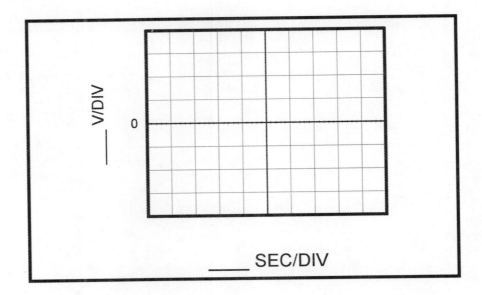

FIGURE 25-3 Amplifier Audio Output Signal with Obvious Clipping

11. Reduce the function generator DC offset voltage to about 1.4 V. Observe the oscilloscope output as you adjust the DC offset. Sketch the resulting output voltage waveform on Figure 25-4.

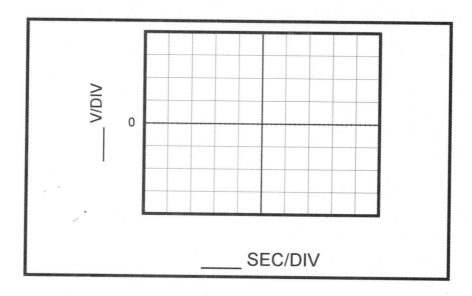

FIGURE 25-4   Amplifier Audio Output Signal with a 1.4-V$_{dc}$ Offset

12. Turn off the signal generator and then turn off the DC supply. Using SW1, connect the dummy load resistor back into your circuit. Reapply the DC power and signal in the proper sequence.

13. Observe the output voltage. It should be similar to the signal observed in Figure 25-4.

Observations

1. Why is it necessary to provide DC biasing for this BJT circuit? Be specific in terms of what the DC biasing must accomplish.

_____

_____

2. Calculate the average signal power and the average total power to the load using your measured RMS values of the output voltage. Why are they different?

_____

_____

_____

_____

Procedure 25-2

BJT Amplifier with Emitter-Supply Biasing

The circuit of Figure 25-1 is very impractical. It is neither reasonable nor typical that the intelligence signal provides biasing (as did the previous circuit with the DC offset in the input signal). Each circuit provides its own biasing to establish its own individual DC Q point about which the signal is to operate. The signal then rides above and below this Q point as it passes through its circuit.

You need to forward bias the transistors Q_1 and Q_2 to turn them "on" and provide enough DC offset to prevent signal clipping. A bias of about 2.5 V_{dc} is needed to provide a 0.7-V drop for each BJT base-emitter junction plus a 1-V offset of the signal to prevent signal clipping.

One solution is to create a separate 2.5-V_{dc} supply that provides a DC base-emitter loop voltage of 2.5 V, which forward biases the BJT base-emitter junction. Another possibility (shown in Figure 25-5) is to bias the emitter side at −2.5 V_{dc} relative to 0 V_{dc} on the base side. This can be accomplished by placing −2.5 V_{dc} at the bottom of the emitter dummy-load resistor and connecting COM to the chassis ground of the function generator. This is called emitter-supply or split-supply bias. This is a very popular bias circuit and is designed to be independent of the transistor beta. The Darlington pair used here also significantly enhances beta independence.

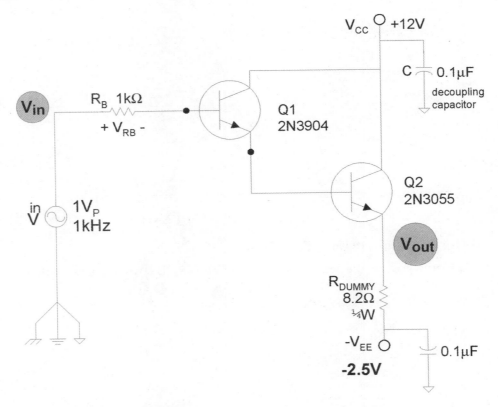

FIGURE 25-5 Emitter Supply Bias −V_{EE}

> ◼ **Note:** Voltage sources ideally provide natural shorts to other sources. Real voltage sources may not act like voltage sources to reverse currents. Thus, care must be taken when sources interact with each other. In this case, the DC bias base current is very, very small and thus should not be a problem flowing through the AC function generator.

1. Construct the circuit of Figure 25-5. First turn on the preset power supplies, set to 12 V and −2.5 V, then turn on the pre-adjusted function generator.

2. Measure the circuit DC voltages and record them in Table 25-3. Do not continue if the measured quiescent DC voltages do not meet expectations. The circuit will not work properly if the DC voltages are not correct.

TABLE 25-3 DC Voltages for the Circuit of Figure 25-5

Quantity	Expected V_{dc}	Measured V_{dc}
V_{CC}		
$-V_{EE}$		
V_{C1}		(STOP)
V_{C2}		(STOP)
V_{B1}		
V_{E1}		(STOP)
V_{B2}		
V_{E2}		(STOP)

(STOP) Instructor sign-off of measured values in Table 25-3 _____

SAMPLE CALCULATIONS

Show your pre-lab calculations for V_E and V_C.

⊐ **Note:** Assume the beta of Q_1 is 200 and beta of Q_2 is 30 for these calculations

$V_E =$

$V_C =$

3. Observe and sketch the total waveform of the signals of Figure 25-5 (scope in DC mode) in Figure 25-6.
4. Are there any differences compared to the preceding observed biased signals?

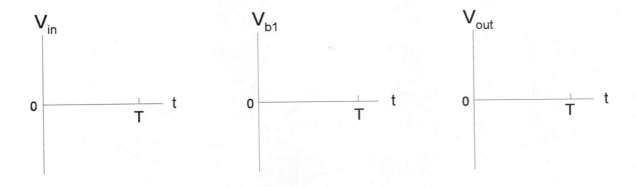

FIGURE 25-6 Emitter Supply of Amplifier Waveforms

🛑 Instructor sign-off of amplifier signals _____

Observations

1. What is the advantage of this circuit over the preceding circuit?

_____

_____

2. Why would a positive DC bias voltage on the emitter side not work?

_____

_____

Procedure 25-3

BJT Amplifier with Voltage Divider Biasing

The emitter supply circuit of the preceding circuit provides a negative DC bias voltage on the emitter side of the BJT.

A dual supply with an additional negative supply is not always available or necessarily desirable. With a single positive supply the circuit needs to produce this 2.5-V_{dc} bias base-emitter loop to forward bias the two base-emitter junctions and provide a 1.0-V_{dc} offset of the signal to prevent clipping.

Another option is to use the available single positive supply voltage and provide a positive DC voltage of +2.5 V_{dc} on the base side of Q_1 relative to common. Like the emitter supply bias circuit, the voltage divider bias circuit is designed to be independent of beta and is therefore very popular. The Darlington pair again significantly enhances the already beta-independent nature of the voltage-divider bias circuit.

⊐ **Note:** Voltage sources ideally provide natural shorts to other sources. An input capacitor must be provided to couple the AC signal (*ideal short*) and block the DC (*ideal open*). Note that this circuit is using an electrolytic capacitor and must be properly inserted with respect to the DC voltage and of sufficient DCWV (DC Working Voltage) rating. The possible result of using an insufficiently rated capacitor or placing it in the circuit backward is that the capacitor may explode! Ensure that your capacitor is sufficiently rated and in the circuit correctly.

SAMPLE CALCULATIONS

Show your pre-lab calculations for your selection of R_2 that will set the base voltage to 2.5 V_{dc}.
Calculated design value for $R_2 =$

---

1. Construct the circuit of Figure 25-7. Connect and turn on the pre-adjusted power supply. Connect and turn on the pre-adjusted function generator.

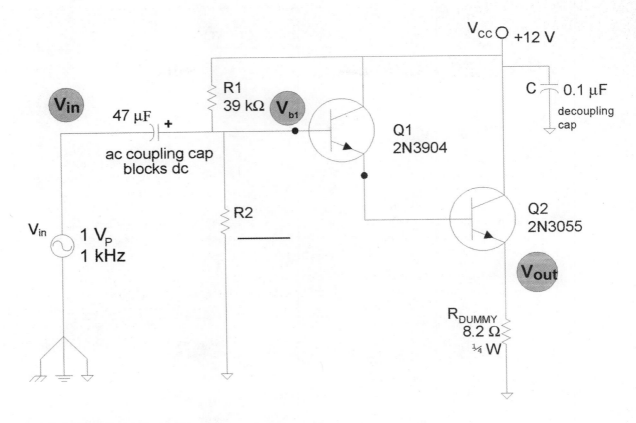

FIGURE 25-7 Voltage Divider Bias

2. Measure and record the circuit DC voltages in Table 25-4.

TABLE 25-4 DC Voltages for the Circuit of Figure 25-6

Quantity	Expected V_{dc}	Measured V_{dc}
V_{CC}		
V_{in}		
V_{C1}		
V_{B1}		
V_{E1}		
V_{C2}		
V_{B2}		
V_{E2}		

SAMPLE CALCULATIONS

Show your pre-lab calculations for:

$V_E =$

$V_C =$

3. Observe and sketch the total waveform of the signals of Figure 25-7 (scope in DC mode) in Figure 25-8. Demonstrate these to your instructor.

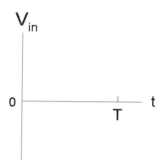
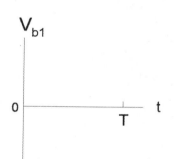
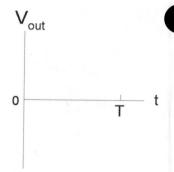

FIGURE 25-8 Voltage Divider Biased Amplifier Signals

🛑 Instructor sign-off of BJT amplifier signals with a resistive bias _____

Observations ◼

1. What is the advantage of this circuit over the preceding circuit?

_____

_____

2. Why would a negative DC bias voltage on the base side not work?

_____

_____

3. What would happen if there were no coupling capacitor on the input signal?

_____

_____

Procedure 25-4

Op-Amp Voltage Follower with a Dual Supply

The operational amplifier is an active device that is composed of several active devices; however, it is treated as a single device with its own unique set of device characteristics. The operational amplifier or op amp is a voltage-controlled, voltage output device. In the circuit of Figure 25-9, the op amp is configured as a voltage follower; that is, the output voltage is the same in amplitude, shape, and phase as its input. This amplifier has a voltage gain of 1 with no phase shift. It amplifies both DC and AC.

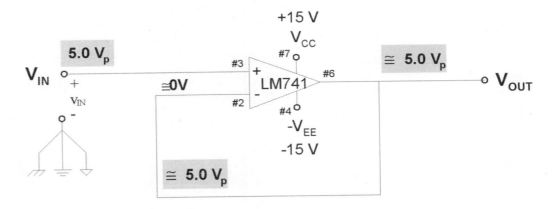

FIGURE 25-9 Operational Amplifier with Dual DC Supplies

The input signal and the output signal have a $0\ V_{dc}$ offset. The output is limited to the rail voltages created by op-amp saturation. Thus, for this circuit the output voltage of $\pm 5\ V_P$ is well within specifications:

$$-15\ V < V_{out} < +15\ V \quad \textit{ideal}\ \text{limits}$$

$$-13\ V < V_{out} < +13\ V \quad \text{more}\ \textit{realistic}\ \text{limits (op-amp saturation)}$$

The dual DC supplies provide the needed voltages to the op amp to properly bias its internal active devices; thus, external biasing is not needed. In fact, neither the input circuit nor the output circuit is required to provide a bias path for the bias current of this circuit. This is much less messy than the preceding cases that mixed the AC and the DC biasing in our external circuit.

The pinout of Figure 25-9 assumes you are using an 8-pin package (see Figure 25-10). If your package is different, then research the data specifications for the pinout of your LM741 op amp.

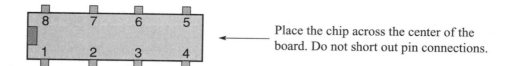

FIGURE 25-10   Pin Connections of an 8-Pin Flat-Pack LM741 Op Amp

1. Construct the circuit of Figure 25-9.
2. Connect the pre-adjusted power supply first. Then apply the pre-adjusted function generator signal. Remember to follow the reverse procedure when powering down (function generator off first, then DC supplies off).
3. Measure and record the DC voltages in Table 25-5.

TABLE 25-5   DC Voltages for the Circuit of Figure 25-9

Quantity	Expected V_{dc}	Measured V_{dc}
V_{CC}	$+15\ V_{dc}$	
$-V_{EE}$	$-15\ V_{dc}$	
V_{in}	$0\ V_{dc}$	
V_{out}	$0\ V_{dc}$	

4. Observe and sketch the total waveform of the signals of Figure 25-9 (scope in DC mode) in Figure 25-11.

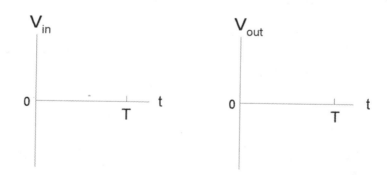

FIGURE 25-11   Op-Amp Voltage Follower Signals

5. Reduce the DC supply voltages until amplitude clipping occurs on both the upper and lower half-cycles. Observe and sketch the input and output waveforms (scope in DC mode) in Figure 25-12. Note the maximum and minimum voltages of V_{out} on your sketch. These are called the rail voltages; the output voltage must lie between these rail values. The rail voltages are typically 80 to 95% of the supplied voltages.
6. Measure the value of the supply voltages at which the clipping occurred.

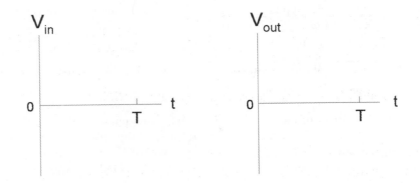

FIGURE 25-12 Voltage Follower with Reduced Power Supply Signals

Observations

1. What is the advantage of this circuit over preceding BJT circuits?

   _____

   _____

2. What are the measured rail voltages for this amplifier? _____ _____

3. What are the measured supply voltages?

   _____

4. What is the value of the rail voltage divided by the supply voltage? _____

Procedure 25-5

Op-Amp Voltage Follower with a Single Supply

Again, a dual supply with an additional negative supply is not always available or necessarily desirable. The preceding operational amplifier circuit utilized a dual DC supply to support the output signal extremes of $\pm 5\ V_p$. However, the circuit of Figure 25-13 only has a single positive supply, which causes a problem.

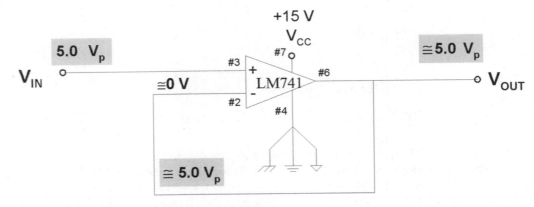

FIGURE 25-13 Operational Amplifier with a Single Supply

With a single positive supply, the circuit needs to produce an output voltage swing of this ± 5 V_P. The rails now become approximately

$$+2 \text{ V} < V_{out} < +13 \text{ V}$$

Since the signal can no longer dip below 0 V, a DC bias of 0 V will cause serious clipping and rectification.

1. Construct the circuit of Figure 25-13. Turn on the pre-adjusted DC power supply first. Then turn on the function generator and increase the amplitude until both half-cycles are clipped. Note the maximum and minimum values of clipping. Is this consistent with the expectations of single source supply of the op amp? These are called the rail voltages. The output must fall between these values. Record the V_{max} and V_{min} values.

   V_{max} due to clipping _____

   V_{min} due to clipping _____

2. Set the function generator amplitude to 5.0 V_P.
3. Measure and record the circuit DC voltages in Table 25-6.

TABLE 25-6   DC Voltages for the Circuit of Figure 25-13

Quantity	Expected V_{dc}	Measured V_{dc}
V_{CC}	$+15$ V_{dc}	
$-V_{EE}$	0 V_{dc}	
V_{in}	0 V_{dc}	
V_{out}	0 V_{dc}	

4. Observe and sketch the total waveform of the signals of Figure 25-13 (scope in DC mode) in Figure 25-14.

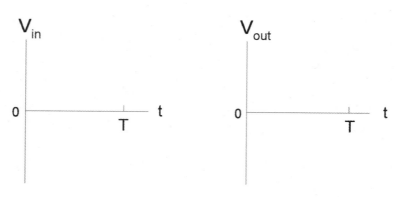

FIGURE 25-14   Single Supply Op-Amp Follower Signals (no biasing)

Observation

1. What is the problem with the output voltage of this circuit?

_____

_____

Synthesis

Op-Amp Voltage Follower with a Single Supply

Use a bias method from a previous procedure to create an op-amp circuit that will operate from a single supply without clipping. You need to raise the input level by a DC offset. You may not add a DC offset to the function generator.

You are limited to the use of a single positive supply of +15 V$_{dc}$. Optimize your design by raising the signal to a DC offset of about one-half of V_{CC}. This permits maximum AC positive- and negative-signal swing on the output without clipping. You may need to use a coupling capacitor to couple the AC and block the DC.

Draw the schematic of your solution on Figure 25-15.

FIGURE 25-15 Single Supply Op-Amp with External Biasing to Prevent Output Clipping (*pre-lab*)

Construct the circuit of Figure 25-15. Apply supply voltages and signals in the proper order, then measure and record the circuit DC voltages in Table 25-7. Add any needed additional DC parameters for your design.

TABLE 25-7 DC Voltages for the Circuit of Figure 25-15

Quantity	Expected V_{dc}	Measured V_{dc}
V_{CC}	+15 V_{dc}	
$-V_{EE}$	0 V_{dc}	
V_{in}		
V_{out}		

Observe the waveforms of the circuit of Figure 25-15 and sketch them on Figure 25-16 (scope in DC mode). Sketch any additional DC nodes you may have created in your design.

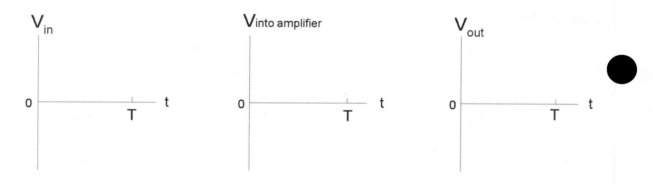

FIGURE 25-16 Single Supply Op-Amp with External Biasing Signals

Increase the function-generator AC amplitude until clipping is observed on both half-cycles. Sketch the clipped signals on Figure 25-17. Demonstrate the clipped signal to your instructor.

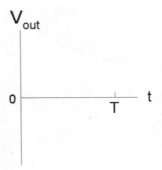

FIGURE 25-17 Single Supply Op-Amp with Clipping on Both Half-Cycles

🛑 Instructor sign-off of clipped amplifier signals _____

Observations ■

1. What is the advantage of this circuit compared to the dual-supply op-amp circuit?

   _____

   _____

2. What is the disadvantage of this circuit compared to the dual-supply op-amp circuit?

   _____

   _____

3. What are the rail voltages for this amplifier? _____ _____

4. Did your designed external bias circuit provide maximum allowable swing of the AC signal without clipping?   ☐ Yes ☐ No

   If not, how would you adjust the bias circuit to improve this situation?_____

   _____

   _____

Special Purpose Analog Integrated Circuit

Name: _____ Date: _____

Lab Section: _____ _____ Lab Instructor: _____
day   time

Text Reference ◼

DC/AC Circuits and Electronics: Principles and Applications
Chapter 20: Special Analog Integrated Circuits

Materials Required ◼

Triple power supply (2 @ 0–20 volts DC; 1 @ 5 volts DC)

1 each 180-Ω and 220-Ω, 1/4-W resistors

2 1-kΩ, 1/4-W resistors

1 1-kΩ single-turn potentiometer

1 each 0.01-μF and 1-μF capacitors

2 0.1-μF capacitors

1 2N3904 bipolar junction transistor

1 XR-2206 function generator integrated circuit

Introduction ◼

Many applications in electronics are met with integrated circuits (ICs) that are designed specifically for a particular task. There are so many special purpose ICs that it would be impossible to list them all.

One such special purpose analog IC is the XR-2206 function generator. This IC can produce square- and sine-wave outputs at a range of frequencies. Additionally, the device will accommodate modulation—a topic not yet discussed. This exercise will explore some of the capabilities of this device. For more information see:

http://www.exar.com/products/XR2206v103.pdf

In this exercise, you will:

▪ Examine the XR-2206 function generator frequency adjustment

▪ Examine the XR-2206 function generator DC offset adjustment

▪ Examine the XR-2206 function generator amplitude adjustment

▪ Examine the XR-2206 function generator VCO operation.

Pre-Lab Activity Checklist

☐ Build the circuit of Figure 26-1.

Performance Checklist

☐ Pre-lab completed?   🛑 Instructor sign-off _____

☐ Table 26-2: Demonstrate measured R_t at 1 kHz.

☐ Table 26-2: Demonstrate measured R_t and C_t at 100 kHz

☐ Table 26-3: Demonstrate measured R_{mult}.

Procedure 26-1

XR-2206 Function Generator Frequency Adjustment

1. Measure and record the value of the 0.01-μF and 1-μF capacitors in Table 26-1.

TABLE 26-1 Measured Capacitor Values

Nominal Value	Measured Value
0.01 μF	
1 μF	

2. Using the 1-μF just measured, apply power to the circuit of Figure 26-1 built in pre-lab.
3. Monitor the square-wave output on pin 11 with the oscilloscope.
4. Adjust the 1-kΩ potentiometer to produce a 1-kHz output frequency.
5. Calculate the value of R_t (180 Ω and potentiometer) resistance needed to provide a 1-kHz frequency. Use the equation: $f_0 = 1/R_tC_t$, where C_t is the measured value of the 1-μF capacitor. Record the value for the calculated R_t in Table 26-2.

TABLE 26-2 Value of R_t and C_t

Frequency	C_t Measured	R_t Calculated	R_t Measured	% Difference
1 kHz				
100 kHz				

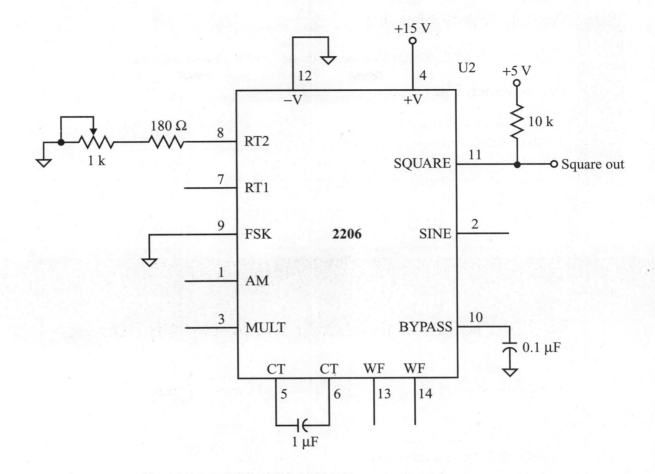

FIGURE 26-1 Function Generator Circuit Using the XR-2206 (*pre-lab*)

6. Remove the power from the circuit. Measure the value of R_t (the series combination of the potentiometer and fixed 180-Ω resistor on pin 8). Record this value as R_t measured in Table 26-2.

7. Calculate the percent difference between the measured and calculated R_t and record the value in Table 26-2.

8. Remove the power from your circuit and replace the 1-μF capacitor at pins 5 and 6 with the 0.01-μf capacitor.

9. Using the measured value of C_t, calculate the theoretical value of R_t needed to produce a 100-kHz output.

10. Restore the power to the circuit and adjust the 1-kΩ potentiometer until there is a 100-kHz signal at pin 11.

11. Remove the power from the circuit and measure the series combination of the 180-Ω resistor and the 1-kΩ potentiometer on pin 8. Record this value, R_t, for the 100-kHz output frequency in Table 26-2.

12. Calculate and record the percent difference between the calculated and measured value of R_t.

13. Return the timing resistors on pin 8 and the capacitors on pins 5 and 6 to the values needed to produce a 1-kHz output. Restore power to your circuit. Verify that the output signal is 1 kHz.

Observations

1. Is your IC an XR-2206M (military grade) or XR-2206CP (commercial grade)?

    ☐ XR-2206M ☐ XR-2206CP

2. Did your IC meet the manufacturer's frequency accuracy specifications at each frequency?

    ☐ Yes ☐ No

Explain your answer: _____

_____

_____

Procedure 26-2

DC Offset with the XR-2206

1. Using the oscilloscope, observe the signal output on pin 2 (triangle/sinusoid output).

2. Remove the power and add the two 1-kΩ resistors and a 10-µF capacitor to pin 3, as shown in Figure 26-2.

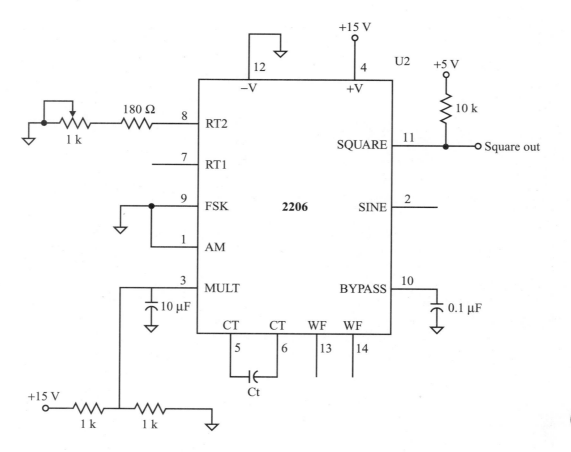

FIGURE 26-2 XR-2206 Function Generator with DC Offset

3. Reapply power and observe the signal output on pin 2.

4. Measure the DC voltage on pin 3 and on pin 2. Record these in Table 26-3.

5. Calculate the percent difference between DC voltage input on pin 3 and the DC output voltage on pin 2. Record your answer in Table 26-3.

TABLE 26-3   DC Offset Voltage

DC Voltage at Pin 3	DC Offset Voltage on Sine-Wave Output (Pin 2)	% Difference

Observations

1. Did the IC perform properly? □ Yes □ No

Why or why not? _____

_____

_____

_____

_____

_____

2. Explain why a DC offset was added to the output.

_____

_____

_____

_____

_____

Procedure 26-3
Amplitude Adjustment of XR-2206

1. Remove power and add the resistor, R_{mult}, to the pin 3 connection, as shown in Figure 26-3.

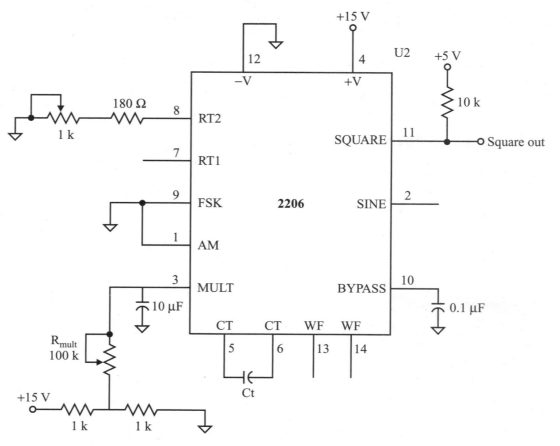

FIGURE 26-3 XR-2206 Function Generator with Amplitude Adjustment

2. Apply power and monitor the output on pin 2 on the oscilloscope.
3. Adjust R_{mult} to produce a 1-V_p output. Record the actual value in Table 26-4.

TABLE 26-4 R_{mult} for Triangle Wave Output		
R_{mult} **Measured**	R_{mult} **Calculated**	**% Difference**